Engineering solids

Engineering Solids

Ian Wilson
Department of Electronic and Electrical Engineering
University of Surrey

McGraw-Hill Book Company (UK) Limited

London · New York · St. Louis · San Francisco · Auckland · Beirut
Bogotá · Düsseldorf · Johannesburg · Lisbon · Lucerne · Madrid
Mexico · Montreal · New Delhi · Panama · Paris · San Juan · São Paulo
Singapore · Sydney · Tokyo · Toronto

Published by
McGRAW-HILL Book Company (UK) Limited
MAIDENHEAD · BERKSHIRE · ENGLAND

British Library Cataloguing in Publication Data
  Wilson, Ian
    Engineering solids.
    1. Solids
    I. Title
    530.4'1'02462   QC176   78-40542

ISBN 0-07-084083-0

1 2 3 4 WC&S 8 0 7 9 8

PRINTED AND BOUND IN GREAT BRITAIN

To Jean, James, Daniel, Edward, Gwilym
and in memory of Brychan

Contents

Preface

This book originated from a course of the same title that is given to first-year electrical engineering students at the University of Surrey. The course serves as an introduction to ideas in areas outside their own narrow topic, but many of the concepts introduced are used later in their physical electronics course. My interest in the subject is based on my experience as a researcher in solid-state physics and also as an industrial metallurgist.

The text has been developed from this nucleus and widened in scope in order to serve as an introduction to aspects of solids of interest to students in other engineering disciplines. It is hoped that the book may, in addition, be valuable as a pre-college introductory text for those intending to study pure science.

The approach has grown out of my own frustrations as a physics student, when I found that one had to become immersed in the details of the subject, the maths, in order to pass exams.

In many cases, the exciting ideas were obscured by the detail until much later. This seems the wrong way of going about things, and this book is an attempt to adopt the opposite approach. The challenge has been to introduce the models without using mathematics, and to integrate the physics with that other important facet of solids, materials science.

The aim was to be concerned solely with ideas, concepts. One may say never mind the width, appreciate the quality. The hope is that if, subsequently, the student comes to measure the width with his mathematical tape measure, he will appreciate what goes into the fabric and how it can be made into a useful garment.

Mathematics has been successfully banished, except for a few relationships and one proof. The proof is included not because it is good for the soul, but to serve as an example of how elegant mathematics can be, and to show how necessary it is in any quantitative description of the physical world.

Integration (of the non-mathematical variety) has not been entirely achieved, the book still falls into two parts: the physics of solids and the science of materials. Perhaps this is the nature of the beast or beasts, but the first two chapters serve as a common base for both parts.

The book is not intended as a reference, to be dipped into to extract facts. *It is meant to be read.* Care was taken to adopt a progressive approach; to tell the story of solids. The simplest model that will explain the phenomenon is always used and every model is developed from one used earlier in the book.

I hope by this means that an interest and understanding of solids will be kindled which, perhaps, will inspire the student to pursue the subject further and in more depth. Even if this aim is not achieved, one hopes, at least, that some insight into how scientists see the physical world has been given, and some of the questions why? and how? have been answered in a simple way.

Finally, I would like to thank Jill Singleton and her colleagues for the enormous help they have given me with the typing and organization of the manuscript. I would also like to thank Ken Stephens and my other colleagues of the Department of Electronic and Electrical Engineering for help and encouragement. For photographs used in the book, I have to thank Jim Whitton of the University of Copenhagen, Peter Goodhew and his colleagues from the Department of Materials Science at the University of Surrey, Peter Hemment of my own department, and the Materials Research Corporation.

Ian Wilson

Part 1

Atomic Bonding and Crystal Structures

'See plastic nature working to this end,
The single atoms each to other tend.
Attract, attracted to, the next in place,
Form'd and impell'd its neighbour to embrace.'

('An Essay on Man', by Alexander Pope)

Part I

Atomic Bonding and Crystal Structures

1. The nature of atoms and the atomic bond

1.1 Introduction

In this book, we will progress from a description of the way atoms congregate to form solids (Part 1) to an introduction to the concepts behind our understanding of the properties of solids (Parts 2 and 3). Finally, Part 4 deals with the ways in which the properties of solids are controlled (one may say engineered) to create materials of use to the technologist.

Models (mathematical or diagrammatical) are used by scientists to describe the physical world that we cannot directly sense, and, therefore, cannot really know in human terms. The models that will be adopted in this book will be the simplest that can be used to describe the phenomena of interest in a non-mathematical way.

We start, in this chapter, with a description of the atomic structure. Once this has been established, the wave nature of sub-atomic particles is discussed by analogy with more familiar vibrating systems (for example, the vibrating string). Using these analogies, the quantization of energy is introduced and the periodic electron configuration of the atom is described. The correspondence between this and the chemical nature of atoms, as evidenced by the periodic table of elements, is then explained.

Armed with this information, we are able to describe the ways in which a bond between two atoms can be formed: the first step in the creation of a solid.

1.2 The atomic structure

The atom is the smallest unit that retains the properties attributable to a particular element. The atom is, therefore, the basic building block of matter.

Rutherford, in 1911, used the bombardment of metal foils with α particles†
to investigate the structure of the atom. Although it was known that the atomic

† An α particle is the nucleus of a helium atom.

3

diameter was of the order of 1 angstrom (1 angstrom is equivalent to 10^{-10} m),† the deflection of these small projectiles by the atoms in a foil told Rutherford that almost all the mass of the atom was concentrated in a region roughly 10^{-4} Å in diameter. This strongly scattering region was called the nucleus of the atom.

The α particles pass right through the majority of the atom without deflection, but something must occupy this space. The reason for saying this is the fact that, when atoms interact to form solids, they arrange themselves in regular arrays, every nucleus being separated from its neighbours by a distance of approximately 1 Å. In fact, the space around the nucleus is occupied by electrons. Our story of solids is mainly that of the electrons; it is they which interact to form the atomic bonds that hold the solid together. In addition, it is the electron configuration that controls the electrical, magnetic, thermal, optical, and mechanical properties of materials.

To introduce atomic structure, we must also consider the nucleus. Therefore, let us split the atom and probe briefly into the world of sub-atomic or elementary particles.

Many elementary particles are now known to inhabit the nucleus; they are involved in creating the nuclear forces. However, it is not the aim of this book to explain the nature of nuclear forces. For a description of atomic structure, we will be safe in making the simplifying assumption that the nucleus is only occupied by two types of elementary particle: the proton and the neutron.

The mass and electrical charge of the three atom-building particles are listed in Table 1.1. We can see that the proton and neutron have approximately the same mass and are about two thousand times heavier than the electron. The nucleus will, therefore, be massive, in agreement with Rutherford's interpretation of the α-particle scattering.

The proton and the electron have equal and opposite charge. This is the smallest unit of electrical charge that can exist, and charge must always be in

Table 1.1   The elementary particles

Name	Approximate mass (atomic mass units)	Charge
Proton	1	$+e$
Neutron	1	0
Electron	0.0005	$-e$

1 atomic mass unit = 1.66×10^{-27} kg.
The charge on an electron, $e = 1.60 \times 10^{-19}$ coulombs

†The angstrom is not a recognized SI unit. However, because it is 'atom sized', it is used in this text for its convenience. It is, in fact, still widely used in many scientific disciplines.

Table 1.2   The first seven isotopes

Name	Symbol	Protons (= electrons), Z	Neutrons N	Mass number, A	Abundance in nature (per cent)
Hydrogen	^{1}H	1	0	1	~100
Deuterium	^{2}H	1	1	2	0.015
Tritium	^{3}H	1	2	3	*
	^{3}He	2	1	3	0.000 13
Helium	^{4}He	2	2	4	~100
	^{6}He	2	4	6	*
	^{8}He	2	6	8	*

*Indicates that the isotope is unstable.

integral amounts of this quantity. This is our first encounter with the 'lumpiness' of properties at an atomic level. We will see later that energy also comes in lumps which are called quanta.

Atoms in equilibrium are electrically neutral; the number of electrons is equal to the number of protons. The number of protons is called the atomic number and is given the symbol Z. It is the value of Z that determines the chemical nature of an atom. When $Z = 1$, we have a hydrogen atom with the chemical symbol H; for $Z = 2$, helium (He); for $Z = 3$, lithium (Li); and so on up to uranium (U) with $Z = 92$, the most massive naturally occurring element.†

You may ask: why do we need neutrons? In fact, neutrons are essential for the stability of the nucleus; it is the interaction between neutrons and protons that produces the enormous force that holds the nucleus together against the electrical repulsion between the positively charged protons. The number of neutrons (N) approximately equals the number of protons (Z) for the light atoms; for heavier atoms, N approaches 1.5 Z.

For any particular value of Z, it is possible to have atoms with different numbers of neutrons. These atoms with the same Z, having therefore an identical number of electrons and an identical chemical nature, differ only with respect to the mass of the nucleus. They are called *isotopes* and are normally represented by writing $^A X$, where X is the chemical symbol (implying a particular value of Z) and A is the total number of particles in the nucleus, i.e., $A = N + Z$; A is called the mass number. The values of Z, N, and A for the isotopes of hydrogen and helium are listed in Table 1.2. Hydrogen is unique in two ways. The most commonly occurring isotope, ^{1}H, has no neutrons in the nucleus (with only one proton, nuclear forces are not needed to stabilize the nucleus). Also, the two other isotopes have their own names and are sometimes given the symbols D and T, although they are chemically identical to hydrogen. Heavy water, which is used in nuclear reactors, is the oxide of deuterium (D_2O instead of H_2O).

†Heavier atoms up to and perhaps beyond $Z = 101$ have been synthesized by man.

Table 1.3 The relationship between atomic number (Z) and mass number (A) for the first 18 elements

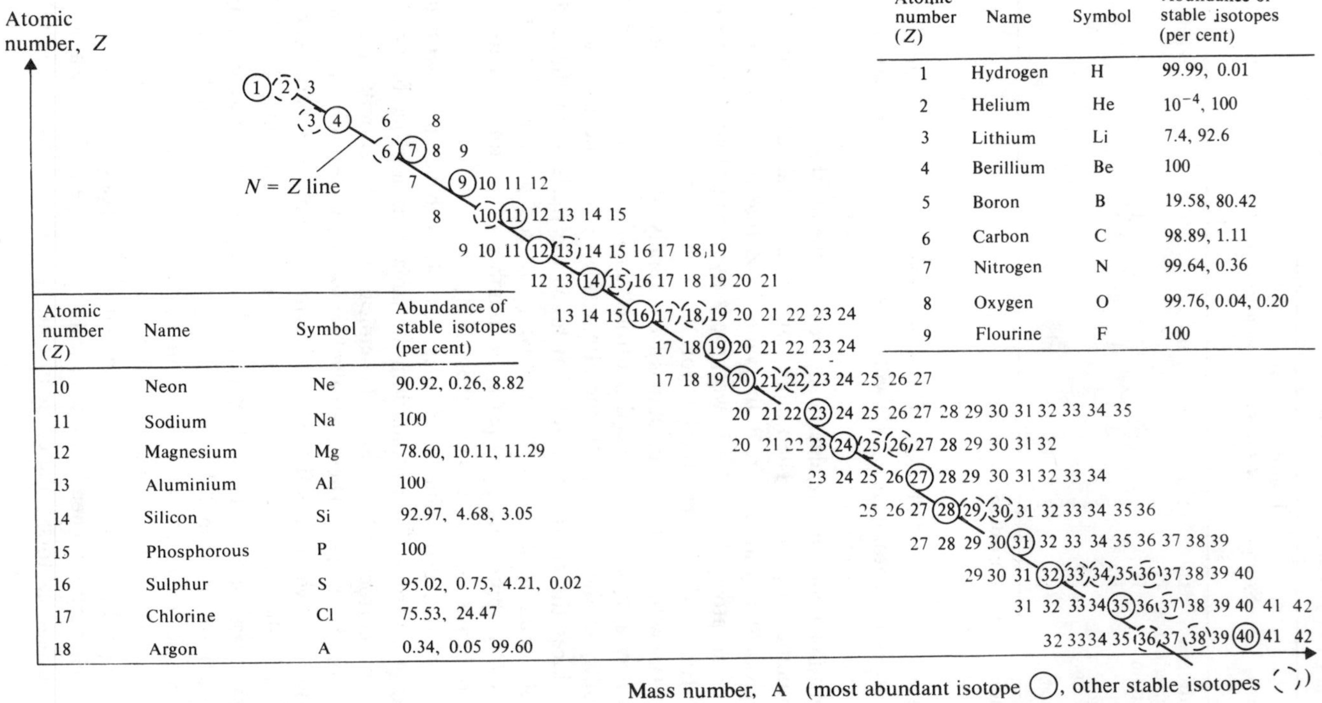

Atomic number (Z)	Name	Symbol	Abundance of stable isotopes (per cent)
1	Hydrogen	H	99.99, 0.01
2	Helium	He	10^{-4}, 100
3	Lithium	Li	7.4, 92.6
4	Berillium	Be	100
5	Boron	B	19.58, 80.42
6	Carbon	C	98.89, 1.11
7	Nitrogen	N	99.64, 0.36
8	Oxygen	O	99.76, 0.04, 0.20
9	Flourine	F	100

Atomic number (Z)	Name	Symbol	Abundance of stable isotopes (per cent)
10	Neon	Ne	90.92, 0.26, 8.82
11	Sodium	Na	100
12	Magnesium	Mg	78.60, 10.11, 11.29
13	Aluminium	Al	100
14	Silicon	Si	92.97, 4.68, 3.05
15	Phosphorous	P	100
16	Sulphur	S	95.02, 0.75, 4.21, 0.02
17	Chlorine	Cl	75.53, 24.47
18	Argon	A	0.34, 0.05 99.60

Atomic number, Z

N = Z line

Mass number, A (most abundant isotope ◯, other stable isotopes ‿)

The horizontal scale shows values of A from 1 to 42. The vertical scale shows values of Z from 1 to 18 together with the name, symbol, and percentage abundance of the stable isotopes.

Three of the isotopes listed are unstable: ^{3}H (tritium), ^{6}He, and ^{8}He. These will decay radioactively, with a release of energy, to form stable nuclei and/or elementary particles.

In Table 1.3, the value of A is plotted as a function of Z for the first 18 elements; the abundance of the stable isotopes is also given. Even for these light nuclei, one can note the tendency for the number of neutrons to exceed the number of protons. One can also see that, for some elements, e.g., chlorine, one stable isotope does not entirely dominate in abundance. This is why the chemical atomic weight, which is an average (weighted by the natural abundance) of the mass numbers of the stable isotopes, does not lie close to an integer. For example, the atomic weight of naturally occurring chlorine is 35.45 a.m.u.

You will probably be familiar with Einstein's famous formula, $E = Mc^2$, where E is energy, M is mass, and c is the velocity of light. This tells us that mass is a form of energy. There are slight mass differences between the sum of the masses of free elementary particles and the mass of a nucleus held together by the strong nuclear forces. The mass difference reflects the magnitude of the binding force. A nucleus is stable if the nuclear mass is less than the mass of the fragments.

The nuclei of massive atoms such as uranium are not stable and they can be induced – by neutron bombardment – to break up (fission) into more stable fragments with a loss in total mass and, therefore, the release of vast amounts of energy. In the nuclear power stations of today, fission of massive nuclei is induced by a well-controlled chain reaction. In the power stations of the future, we may harness nuclear energy by the fusion of unstable light isotopes (such as tritium) to create more stable isotopes (such as ^{4}He), again with a reduction in total mass. We know it can be done because we see a natural fusion reactor nearly every day – the sun.

This is as far as we need to go in description of the nucleus. We will now turn our attention to the electrons occupying the space around the nucleus. As mentioned earlier, the electrons are of fundamental importance in understanding interatomic bonding and, therefore, the structure and properties of solids. They are also the source of chemical energy, where power is generated by burning wood, oil, gas, or coal. Again, energy is derived by a reduction in mass. In this case, the reduction in mass occurs when more stable interatomic bonds are formed. The mass change, and, therefore, the energy produced, is very much smaller than for the nucleus. This is because the electron binding energy is very much less than the nuclear binding energy.

1.3 Waves and particles

The model that we will use for an electron is *not* the one that is popularly shown of a small particle whizzing around the nucleus like a satellite orbiting a planet.

The electron and all other atomic-sized 'particles' have many of the properties

Figure 1.1 Electron diffraction pattern created by the passage of a beam of electrons through a thin crystal of gold.

that are attributed to waves. This can be readily demonstrated in the electron microscope, where a focused beam of electrons can be diffracted by the regular arrangement of atoms in a single crystal. This produces a pattern such as that shown in Fig. 1.1. This is called a diffraction pattern and arises from the same phenomenon as the formation of a spectrum when white light is incident on an optical grating. The diffraction pattern is caused by constructive interference of scattered waves.

The process of diffraction is illustrated in Fig. 1.2. The electron is represented by a continuous sine wave of wavelength λ. This sine wave is incident on a crystal lattice. The atoms scatter the waves in all directions, but the scattered waves interfere destructively unless the waves scattered from *all* the atoms are exactly in phase. This will occur for only two situations: first, when the transmitted wave remains in phase and continues as though not scattered (although it may be diminished in amplitude); second, when, as illustrated, the diffracted wave scattered from one atom is in phase with that scattered from a neighbour because the extra path length down to the neighbour ($d \sin \theta$) and back up from the neighbour (again, $d \sin \theta$) is equal to an integral number of wavelengths ($n\lambda$). This leads to the well-known Bragg diffraction relationship, $2d \sin \theta = n\lambda$, where d is the separation between neighbouring atomic planes in the crystal, θ is the angle between the direction of incidence of the wave and the atomic plane, and λ is the wavelength of the incident radiation (X-rays or

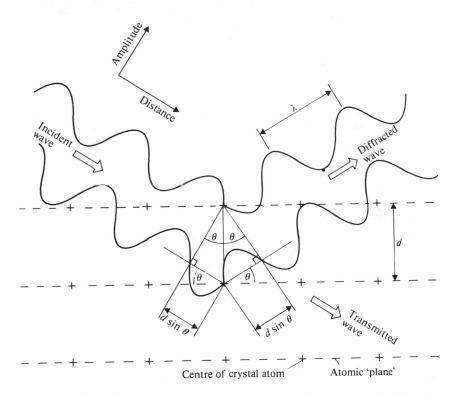

Figure 1.2 Schematic representation to illustrate the conditions necessary for diffraction of an electron wave by a crystal.

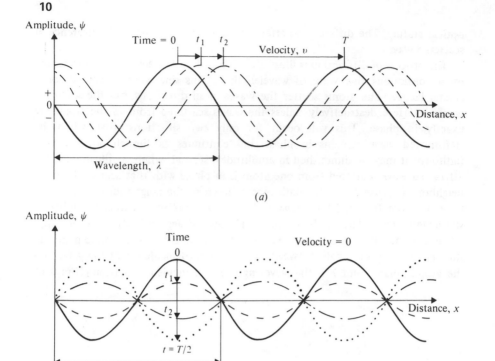

Figure 1.3 One-dimensional waves. (*a*) Travelling wave; (*b*) standing wave. Broken curves represent the amplitude at different times.

electrons). The scattering is as though the atomic plane were a mirror, reflecting the waves with the angle of incidence equal to the angle of reflection.

Each spot and circle on the electron diffraction pattern in Fig. 1.1 represents scattering from a particular atomic plane in a thin crystal of gold. As we need to know something of the properties of waves in order to discuss the behaviour of electrons in solids, a little time will be spent on this subject.

We can consider waves to be of two general types, illustrated in Fig. 1.3. The wave shown in Fig. 1.3(*a*) is a travelling wave. The illustration represents the wave amplitude as a function of distance at various times. This is a one-dimensional wave such as one would set up by vibrating the end of a rope of infinite length. Travelling waves generally exist in more than one dimension; for example, ripples caused by throwing a pebble in a pond, electromagnetic radiation such as light, sound waves from a speaker, the electrons in an electron microscope, and even the electrons that carry electrical current in solids.

The wave shown in Fig. 1.3(*b*) is a standing wave, and represents a resonant condition caused by the superposition of waves travelling in opposite directions.

This will occur when the end of a rope of finite length is vibrated. A practical example is the violin string. The travelling wave launched in one direction interferes with one launched earlier that has been reflected from the end; the interference is only constructive at the resonant frequency. Examples of standing waves are the vibrations of the sound box of a stringed instrument, the vibrations in an electronic oscillator, and the vibrations of an electron that is bound to a nucleus in an atom.

We will come back to standing waves when discussing the energy of electrons in atoms. First, however, we will consider the travelling wave, by which energy can be transmitted from one place to another. The transfer of energy can be illustrated by considering a twig floating in a pond. If a pebble is thrown into the pond, the twig is caused to bob up and down. Energy is transmitted to the twig from the point of pebble impact by travelling waves.

Imagine again the rope of infinite length in Fig. 1.3(a). Let us say that it takes time T to move the rope end (in simple harmonic motion) up from the central position to maximum amplitude up, then down through the central position to maximum amplitude down, then back to the central position. We will have produced one wave of length λ in that time. Thus, the wave has travelled a distance λ in time T. The velocity of the wave is, therefore, given by $v = \lambda/T$ or $v = \nu\lambda$, where ν is the frequency of the wave (the number of waves passing per unit time) given by $\nu = 1/T$.

We are familiar with the idea of light as a travelling wave with a velocity of 3×10^8 m s^{-1}. The mean wavelength in the visible spectrum is about 5000 Å (or 0.5 μm). The frequency of visible light is, therefore, about 10^{14} Hz (1 hertz = 1 cycle per second). You may be familiar with the tuning dial on your radio receiver, calibrated in megacycles or megahertz ($\nu = 10^6$ Hz). As radio waves also travel at the velocity of light, it is easy to work out that megahertz radio waves have wavelengths in the hundreds of metres. X-rays, on the other hand, have wavelengths in the angstrom range and are strongly scattered by atoms.

Waves are not disturbed by objects which are small compared with their wavelength, so that visible light can travel through the atmosphere,† and through insulating solids such as glass, without being scattered by the atoms. The velocity is reduced in a dense medium and this phenomenon leads to the most wave-like property of light: refraction. It is the refraction of light by glass that enables us to make optical lenses.

Light cannot travel through electrical conductors because the light energy is absorbed by the free electrons that exist in abundance in conducting solids. The free electrons, in fact, can receive so much energy from the light that they are ejected from the surface. This is known as the photo-electric effect. It has the curious feature that the ejected electrons come off with energies that depend

---

†The larger molecules in the atmosphere do scatter light to some extent, especially the shorter (blue) wavelengths, which is why the sky is blue.

12

Table 1.4   Derivation of the de Broglie relationship between wavelength and momentum of a wave/particle

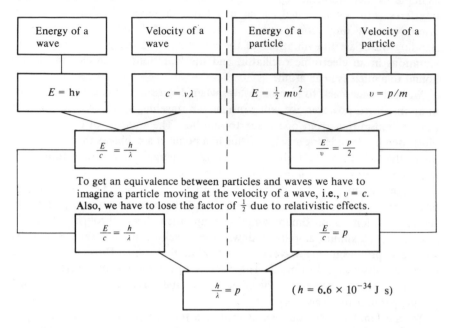

only on the *frequency* of the wave and not, as one would expect of a continuous wave, on the *amplitude*. This can only be explained if one assumes that light comes in packets and every packet gives up its entire energy to one electron.

We run into the wave/particle duality of matter when we get down to the atomic scale of things.

Planck was the first person to establish the particle-like properties of light when he sought to explain the spectrum of radiation emitted from hot bodies. Planck had to assume that light came in units of energy, $E = h\nu$, where h is a constant $(6.6 \times 10^{-34}$ J s$)$. The unit, packet, or 'particle' of light was later named the photon. The photon is a strange particle in that it has no mass and no charge. In fact, Einstein's theory of relativity states that the photon *must* have zero mass in order to travel at the speed of light.

One can use Planck's expression for the photon energy together with the expression for the kinetic energy of a particle with mass m to derive a relationship between the particle-like property of momentum $p = mv$ and wavelength λ. This is done in Table 1.4. We have to make the unreasonable assumption that the 'particle' is travelling at the speed of light and pay for this by losing the factor of $\frac{1}{2}$ from the kinetic energy expression. Although the derivation in Table 1.4 is not rigorous, the result, the de Broglie relationship $p = h/\lambda$, certainly holds true. It was de Broglie who first appreciated that not only does light have particle-like properties, but also that nuclear particles have wave-like properties.

1.4 Wave packets and uncertainty

How can we visualize the atomic wave/particle? We can again use the analogy of an infinite rope held taut. If, as illustrated in Fig. 1.4, the rope end is shaken violently for a short time, a wave train or wave packet is launched and travels along the rope. The wave packet may be composed of waves of several wavelengths, but it travels along the rope at a lower velocity (the group velocity) than that of the component waves. The amplitude is zero everywhere except in the wave packet. It was established by Fourier (1768–1830), well before modern physics, that a wave train of finite length is equivalent to the sum of many infinite constant-amplitude waves of different wavelengths. Hi-fi experts will know that one requires amplifiers with a very large dynamic range (from very low frequencies to frequencies well above the audible limit) in order to reproduce very short percussion sounds accurately.

The effect of the superposition of only two travelling waves is illustrated in Fig. 1.5. The waves superimpose to create a train of 'beats', with maximum amplitude where the waves coincide exactly. The width of a beat (Δx) between

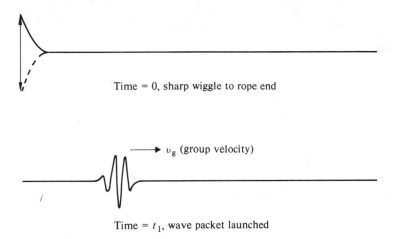

Time = 0, sharp wiggle to rope end

v_g (group velocity)

Time = t_1, wave packet launched

v_g

Time = t_2, wave packet continuing

Figure 1.4   Transmission of a wave packet along a stretched rope.

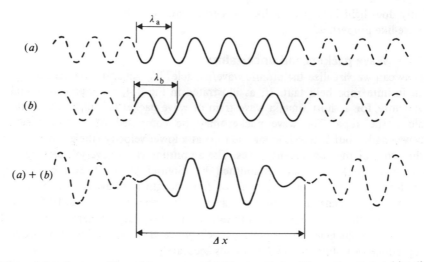

Figure 1.5 Superposition of two waves. (*a*) Wavelength λ_a; (*b*) wavelength λ_b; (*a*) + (*b*) beats formed of width Δx.

the points of zero amplitude is such that, in this example, one wave (λ_a) has completed five cycles and the other (λ_b) has completed four cycles. So, $\Delta x/\lambda_a = 5$ and $\Delta x/\lambda_b = 4$; therefore,

$$\Delta x\left(\frac{1}{\lambda_a} - \frac{1}{\lambda_b}\right) = 1.\dagger$$

If we change wavelength into momentum using the de Broglie relationship, then

$$\Delta x\left(\frac{p_a}{h} - \frac{p_b}{h}\right) = 1.$$

Let the difference in momentum $p_a - p_b = \Delta p$; then,

$$\Delta x \Delta p = h.$$

This combination of uncertainties in position and momentum is identical to the prediction of the Heisenberg uncertainty principle, which expresses in another way the consequence of the wave/particle duality of matter:

one can never know exactly, at one instant of time, both the position and the momentum of a wave/particle.

Take the case of a single infinite wave. The wavelength is exactly defined; therefore, $\Delta p = 0$. However, as the amplitude is constant throughout space and

†This holds generally true for beats, where $\Delta x = n\lambda_1 = (n + 1)\lambda_2$.

time, we have no information of the position of the wave/particle and, therefore, the uncertainty in position Δx is infinite.

If two such waves of different wavelength are superimposed, we have the situation shown in Fig. 1.5. However, our real knowledge of position is still poor, because we have an infinite train of beats, each of width Δx. If more than two wavelengths are superimposed, they can usually only add constructively at one point; thus, as more and more waves of different wavelength are superimposed, one beat will increase in amplitude at the expense of the rest. Also, Δx, the width of the beat, will get less and less as the number of wavelengths and, therefore, Δp gets greater.

We can now see the atomic wave/particle as a wave with non-zero amplitude only in a small region of space, Δx.† This can only be achieved by the superimposition of waves with a range of wavelengths, which implies that there is an uncertainty in momentum, Δp. A free electron, for example, acts in this way as a superposition of travelling waves to create a travelling wave packet. The electron bound to an atom is quite a different case and is much more analogous to a standing resonant wave.

1.5 Standing waves and quantization of energy

The fact that a bound electron is a wave confined to the small volume around an atom has important consequences when we consider the energy of the electron. We find that the electron is not free to have any value of energy; rather, there is a discrete set of allowed values of energy. On an atomic scale, energy comes in units called quanta, and we describe the energy of an electron by saying that it is in a particular stationary state (stationary because a standing wave has zero group velocity). Quantization of energy is a direct result of the wave nature of electrons.

As we have stated, each electron and each proton has one unit of charge, and this is indivisible. The quantum of energy is somewhat different, in that it can vary in size; the important concept is that of a discrete set of stationary states or energy levels into which we can fit the number of electrons required to make the atom neutral. The electron is a complex mixture of waves existing in three dimensions and it is impossible to demonstrate here why confinement to a small volume of space infers quantization of energy. However, we can take a simple one-dimensional example to illustrate the effects of putting bounds on a vibrating system. The example we will use is the violin string.

The velocity of propagation of a wave along a stretched string depends on the tension and linear density of the string; these factors, together with length, will determine the frequency of vibration, i.e., the musical note. We are concerned here only with the *wavelengths* of the resonant modes that occur when reflected waves interfere constructively with transmitted waves.

† Δx is of the order of the fundamental wavelength.

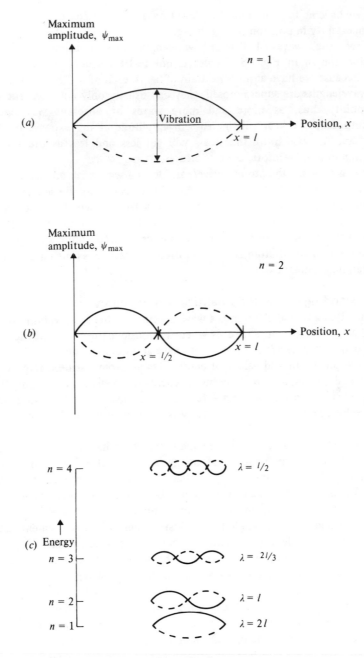

Figure 1.6 Modes of vibration and energy levels for a violin string. (*a*) Mode for $n = 1$; (*b*) mode for $n = 2$; (*c*) energy levels and modes for $n = 1$ to 4.

The boundary condition that determines the resonant frequencies of this system is that the string has a fixed length l. Figure 1.6 illustrates the positions of the string at the instant of maximum amplitude for the two lowest energy (longest wavelength) resonances: state numbers $n = 1$ and $n = 2$. The resonances occur at $\lambda = 2l$ for $n = 1$, and $\lambda = l$ for $n = 2$. The third resonance ($n = 3$) will occur at $\lambda = 2l/3$ and the nth resonance at $\lambda = 2l/n$.

The nature of the energy of a vibrating string changes from 100 per cent potential energy (tension), at the moment of maximum amplitude, to 100 per cent kinetic energy when the string is instantaneously straight. If we take this latter case, we can work out the relative energy of the stationary states:

Kinetic energy, $E = mv^2/2 = p^2/2m$.

But we know from the de Broglie relationship that $p = h/\lambda$ and, from above, $\lambda = 2l/n$. Therefore, $p = hn/2l$ and

$$E_n = (h^2/8ml^2)n^2.$$

The energy of the nth mode E_n is proportional to the square of the state number for a vibrating string. This is illustrated in Fig. 1.6 with an energy-level diagram, each level labelled with the state number.

The electron bound to an atom is also a resonant system, with energy constantly changing from potential energy to kinetic energy, and with only discrete values of total energy allowed. One can ascribe a quantum number to each state, but the energy is not proportional to n^2; in fact, the levels get closer together as n increases.

This difference arises from the fact that the restoring force for a violin string is described by Hooke's law: force is proportional to displacement. The electron restoring force is the inverse square law of Coulombic attraction between opposite charges.

The stationary states for an electron bound to a nucleus by Coulombic attraction have been worked out by Schrödinger, assuming that the electron is a wave. For this, a new branch of mathematical physics, called wave mechanics, had to be evolved. This is too complex for us to describe here, but it predicts stationary states that are analogous to the violin-string modes.

An earlier semi-classical model of Bohr yields a result that is strikingly similar to that for the vibrating string. Bohr assumed that the electrons are particles in orbit around the nucleus, like satellites orbiting a planet, but used some of the ideas of quantum theory in that he assumed that only certain orbits were stable, that electrons could only make quantum jumps of energy $E = h\nu$, and that angular momentum was quantized (more of this last phenomenon in Part 2).

The result of the Bohr model is that the electron energy is given by

$$E_n = \left(\frac{h^2}{8m(\pi r)^2} \right) \frac{1}{n^2}.$$

This is strikingly similar to the violin string solution, with l replaced by the electron 'orbit' radius r times π. The energy is now dependent on the reciprocal of n^2, where n is called the principal (or total) quantum number. Each electron bound to an atom must occupy a stationary state or energy level characterized by a set of *four* quantum numbers. The violin-string model only helps us to understand the principal quantum number, which is the most important of the four in determining the energy of the electron.

1.6 The names of electrons

We will now visualize the atom as a bare nucleus with a set of possible stationary states for electrons, each with its own unique set of four quantum numbers and, therefore, its own energy level. Which states will be occupied when we add electrons to make the atom neutral? We would first say that, as nature conserves energy, they would all occupy the lowest energy state, the fundamental, of quantum number $n = 1$; but, in fact, this is impossible. No two electrons that inhabit the same atom can have the same energy, i.e., the same set of quantum numbers. This is called the *Pauli exclusion principle*; we will invoke it again in Part 2, when we deal with the ways in which stationary states interact when we form large assemblies of atoms. This principle applies to any system of interacting electrons.

Our next guess may, therefore, be that the first electron would go into the lowest energy state $n = 1$, the second into $n = 2$, and so on; again, we would be wrong. Because of the complex way in which the electron vibrates, we have several *types* of vibration, each with its own quantum number for each value of the principal quantum number. Traditionally, these are called s, p, d, and f states or 'orbitals'. These increase slightly in energy in the order given as a fine structure on the energy-level scheme. There are two more quantum numbers, arising from the magnetic moment of the electron and the magnetic moment of the stationary state as a whole. These only denote a different energy when the atom is in a magnetic field, but imply that more than one electron can occupy each s, p, d, or f state.

The details of the four quantum numbers will be given in Part 2. For now, all we need in order to characterize or name each electron attached to an atom is the principal quantum number (integers 1 to 7) and the type of state (s, p, d, or f). The distribution of electrons among the allowed stationary states is governed by the following rules.

1. $n = 1$, only s states;
   $n = 2$, s and p;
   $n = 3$, s, p, and d;
   $n = 4$, s, p, d, and f;
   $n > 4$, we still have s, p, d and f. (Not all possible states are usable because nuclei become too large and are unstable.)

2. Each *s* state can have up to two electrons;
   each *p* state can have up to six electrons;
   each *d* state can have up to 10 electrons;
   each *f* state can have up to 14 electrons.
3. The *d* states fill one principal quantum number late because their energy is greater than the next *s* state up, i.e., the *s* states for $n = 4$ (4*s* states) fill before 3*d*, 5*s* states fill before 4*d*, and so on. *f* states fill two principal quantum numbers late, so 6*s* fill before 4*f* and 7*s* fill before 5*f*.

The energy level scheme for electrons on an atom, based on the quantum number rules, is shown in Table 1.5. The heavy lines enclose the electrons in a particular period or 'shell'. The question marks indicate that further massive atoms may be created artificially, but these would be pretty unstable because of the nature of forces holding the nucleus together.

Using the set of rules listed above, we will now assemble the series of elements by adding electrons until their number is equal to that of the protons in the nucleus (neutrons play no part in this game).

If the nucleus has a single proton, we need just one electron. This can go into a 1*s* state to make an atom of hydrogen. If we need a second electron, this can

Table 1.5 **The number of electrons in stationary states for all the elements (principal quantum number n = 1 to 7, orbital names *s*, *p*, *d*, and *f*)**

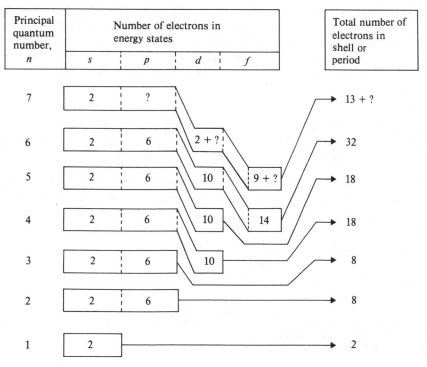

also go into the $1s$ state and we would have an atom of helium. The electron configuration (or name) of an atom of helium is written as $1s^2$, where the superscript shows the number of electrons in the $1s$ state. Now, all the states for $n = 1$ are filled and we have what is termed *a closed shell*. If another electron is required, it would have to go into a $2s$ state (lithium), the next can also go into $2s$ (beryllium), but the next (to make an atom of boron) would have to go into a $2p$ state. The $2p$ state is filling as we progress from boron through carbon, nitrogen, oxygen, and fluorine until neon $(1s^2 2s^2 2p^6)$ is reached, where all the states for $n = 2$ are filled and we have once again a closed shell. The next closed shell is for the element argon $(1s^2 2s^2 p^6 3s^2 3p^6)$; you may notice that this occurs *before* the $3d$ states start to fill. These fill one principal quantum number late, so the next closed shell is krypton, with the configuration $1s^2 2s^2 2p^6 3s^2 3p^6 4s^2 3d^{10} 4p^6$.

We have used the concepts of modern physics to describe the classification of the elements, but chemists already knew this classification from chemical reactivity. They call this energy-level scheme the periodic table.

1.7   The periodic table

The periodic table is shown in Table 1.6, where the elements are shown by their chemical symbol. The number of electrons (the atomic number Z) is shown for each element.

Chemical reactions have long demonstrated that it is the number of electrons in the incomplete shell that determines the type of atomic bonding and, thus, the chemical nature of the atom. The electrons in partially filled states are called the valence electrons, and it is these that take part in the formation of bonds.†

The clearest examples come from the 'main line' part at the top of Table 1.6, where the s and p states are filling, namely:

closed shell (Group 0)   He, Ne, Ar, Kr, Xe, Rn,   the inert gases;
one extra electron (Group la)   Li, Na, K, Rb, Cs,   the alkali metals;
one electron short of a closed shell (Group 7a)   F, Cl, Br, I,   the halogens
four electrons (Group 4a)   C, Si, Ge,   the semiconductors.‡

These are the obvious groupings, and this is where chemistry and physics come together, because the chemical nature of an atom arises from the electron distribution. The electron distribution, in turn, is determined by the set of quantum numbers describing the stationary states.

There are more subtle relationships between the elements, and chemists spend happy hours scanning the periodic table for diagonal relationships and so on. Some important trends can be mentioned here.

---

†The valency is not always equal to the number of valence electrons; for example, chlorine can act as though there is an excess of one or five electrons, as well as the excess of seven or lack of one that we would expect.
‡Carbon as diamond.

Table 1.6   The periodic table of the elements

Electron occupancy / Group number	s states filling		p states filling					All states filled	Names of outer electrons
	s^1	s^2	s^2p^1	s^2p^2	s^2p^3	s^2p^4	s^2p^5	s^2p^6	
	1a	2a	3a	4a	5a	6a	7a	0	
	H^1							He^2	$1s$
	Li^3	Be^4	B^5	C^6	N^7	O^8	F^9	Ne^{10}	$2s\,2p$
	Na^{11}	Mg^{12}	Al^{13}	Si^{14}	P^{15}	S^{16}	Cl^{17}	Ar^{18}	$3s\,3p$
	K^{19}	Ca^{20} *	Ga^{31}	Ge^{32}	As^{33}	Se^{34}	Br^{35}	Kr^{36}	$4s\,4p$
	Rb^{37}	Sr^{38} **	In^{49}	Sn^{50}	Sb^{51}	Te^{52}	I^{53}	Xe^{54}	$5s\,5p$
	Cs^{55}	Ba^{56} ***	Tl^{81}	Pb^{82}	Bi^{83}	Po^{84}	At^{85}	Rh^{86}	$6s\,6p$
	Fr^{87}	Ra^{88}							$7s$

Group number	d states filling (transition elements)										Names of outer electrons
	3b	4b	5b	6b	7b	8			1b	2b	
*	Sc^{21}	Ti^{22}	V^{23}	Cr^{24}	Mn^{25}	Fe^{26}	Co^{27}	Ni^{28}	Cu^{29}	Zn^{30}	$3d\,4s$
*	Y^{39}	Zr^{40}	Nb^{41}	Mo^{42}	Tc^{43}	Ru^{44}	Rh^{45}	Pd^{46}	Ag^{47}	Cd^{48}	$4d\,5s$
*	La^{57} †	Hf^{72}	Ta^{73}	W^{74}	Re^{75}	Os^{76}	Ir^{77}	Pt^{78}	Au^{79}	Hg^{80}	$5d\,6s$
*	Ac^{89} ‡										$6d\,7s$

	f states filling	Names of outer electrons
†	Lanthenides (rare earths) from Ce (58) to Lu (71) including Gd (64)	$4f\,6s$ ($5d$ sometimes)
‡	Actinides from Th (90) to Lr (103) including U (92) and Pu (94)	$5f\,7s$ ($6d$ sometimes)

1. The few non-metals (which include the all important life-giving elements, carbon, oxygen, and nitrogen) occupy the upper right-hand side of the periodic table, with not many electrons and almost full shells.
2. The three ferromagnetic elements, iron (Fe), cobalt (Co), and nickel (Ni) occupy a block comprising Group 8 of the first row of transition elements (d states filling).

Table 1.7   Some properties of the elements in the third row of the periodic table

	Group							
	1a	2a	3a	4a	5a	6a	7a	0
Chemical symbol	Na	Mg	Al	Si	P	S	Cl	Ar
Atomic number	11	12	13	14	15	16	17	18
Outer shell electrons	$3s^1$	$3s^2$	$3s^2 3p^1$	$3s^2 3p^2$	$3s^2 3p^3$	$3s^2 3p^4$	$3s^2 3p^5$	$3s^2 3p^6$
Crystal structure†	b.c.c.	h.c.p.	f.c.c.	Diamond		Complex		f.c.c.
Electrical resistivity at room temperature‡	5.2	4.45 Metals	2.65	2.5×10^5 Semi-con-ductor	1×10^{17} Insulators	2×10^{23}	Normally gas, conductivity of solid unknown	—
Bond strength§	1.13	1.53	3.34	4.62	?	2.86	?	0.080
Melting point (°C)	98	650	660	1400	44	120	−100	−189
Boiling point (°C)	890	1100	2500	2400	280	440	−35	−186

†b.c.c., body-centred cubic; f.c.c., face-centred cubic; h.c.p., hexagonal close-packed.
‡In units of 10^{-4} Ω m. Copper is 1.67×10^{-4} Ω m. The silicon value is the 'intrinsic' resistivity.
§Energy in electron volts per atom to separate two neutral atoms. An electron volt is the energy an electron would acquire when accelerated by a potential difference of 1 V.
1 eV = 1.6×10^{-19} J and is equivalent to a temperature of 10^4 K.

3. The eight noble† metals occupy a block on two periods just below this.
4. The energy levels are very close together when the d and f states are filling, so there are irregularities in the filling order. For example, the valence electron configurations of iron, cobalt, and nickel are, respectively, $3d^6 4s^2$, $3d^7 4s^2$, and $3d^8 4s^2$, whilst the next element, copper (Cu), is $3d^{10} 4s^1$. The other two good electrical conductors, silver (Ag) and gold (Au), have similar valence electron configurations ($4d^{10} 5s^1$ and $5d^{10} 6s^1$).

Our model for the atom is becoming rather complicated, but the complications of d and f states have only been included for completeness (although we will have to consider d states when talking about ferromagnetism in Chap. 4).

We need only to consider a simple period in order to arrive at a model for the four important types of atomic bonding. We will take as our example the third period ($n = 3$), where the atomic structure is a core of neon plus the valence electrons.‡

†Ru, Rh, Pd, Ag, Os, Ir, Pt, Au.
‡In the same period are examples of the four most common crystal structures to be described in Chap. 2.

Some properties of these eight elements are listed in Table 1.7. The elements follow the sequence, metal, semiconductor, insulator, gas, as the number of valence electrons increases. Bond strength (and, therefore, melting and boiling points) peaks at the centre of the period, in Group 4a, where there are four valence electrons.

These properties derive from the nature of the interatomic bonding, which, in turn, is determined by the valence electron configuration. We now have a sufficiently sophisticated model of the atom to describe the electron interactions that give rise to atomic bonding.

1.8 Interatomic bonds

As we bring two atoms together and their separation decreases, the electrons begin to interact; this interaction is such that the electrons can almost always rearrange themselves and reduce the total potential energy in our two-atom system. Because nature always tries to reach the lowest energy configuration possible, there is an attractive force between the two atoms. As illustrated in

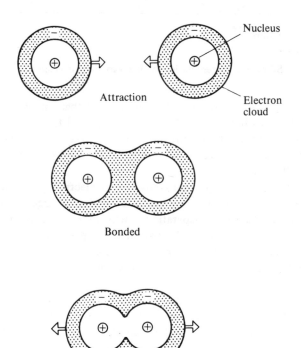

Figure 1.7 Attraction, bonding, and repulsion using a simplified model of an atom.

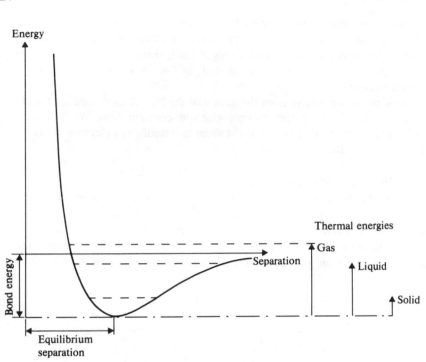

Figure 1.8 Schematic representation of the variation of electron potential energy with atomic separation.

Fig. 1.7, once the electron clouds overlap more than an optimum amount, the 'veil' of electrons between nuclei becomes thin and the nuclei repel each other very strongly due to the electrostatic field between them. This repulsion is so strong that physicists refer to some atomic collisions as 'hard-sphere' or 'billiard-ball' collisions.

In Fig. 1.8, a schematic representation of electron potential energy against atomic separation is plotted as a gentle attraction at large separations followed by a strong repulsion. The separation at which the energy is a minimum is called the equilibrium separation, and a pair at this separation are said to be bonded. The electronic configuration between atoms, i.e., the type of bonding, determines the form of all matter as we know it on earth, and, presumably, on other planets.

One important state of matter which we will not deal with here is not solid, liquid, or gas and is not composed of atoms at all. This is the hot, but electrically neutral, soup of sub-atomic particles that makes up the stars; this state of matter is called plasma.

We will use Fig. 1.8 to differentiate between the three terrestrial states of matter by introducing the concept of thermal energy. Atoms usually have some kinetic energy; they are moving all the time – exchanging energy with each

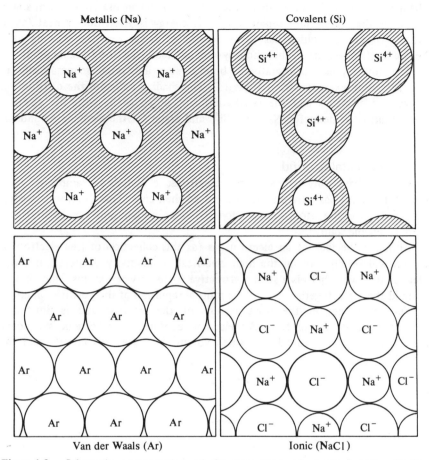

Metallic (Na)   Covalent (Si)

Van der Waals (Ar)   Ionic (NaCl)

Figure 1.9 Schematic representation of the four types of atomic bonding. Shading represents the valence electrons. Circles represent closed shells.

other. The subject of heat will be dealt with in detail in Chap. 6; however, one can see that, if the thermal energy of an atom is much greater than the electron energy saved in creating a bond, then no stable bonds will form and the atoms will be almost free of influence from each other. In this case, we will have a gas.

If, on the other hand, the thermal energy is small compared with the bond energy, then stable bonds will form and atoms will stay at the same average position with respect to one another, oscillating gently about this position. This is a solid.

Liquids are the intermediate case, there is not enough thermal energy to 'break' the bonds completely, but atoms can take up a wide range of positions with respect to one another.

Latent heat can now be understood. This is the energy saved, for example, on changing from the liquid state to the solid state, when the atoms form stable bonds at the equilibrium separation. This energy is given out as heat (extra atomic vibrations), which has to be conducted away from the growing solid before the temperature of the whole system can fall below the freezing point.

The four types of interatomic bonding are illustrated schematically in Fig. 1.9, using examples from Table 1.7. The shading indicates the distribution of valence electrons and the circles indicate closed shells with the electron configuration of neon for Na^+ and Si^{4+} and argon for the element Ar and for Cl^-.

1.9 The metallic bond

The metals have only a few valence electrons; in fact, the example we have chosen (sodium) has only one, with a closed shell core. This electron is readily detached once the atoms come close to one another and can roam freely throughout the solid.

Although the idea of an electron gas (as this collection of free electrons is called) will readily explain the high electrical conductivity of metals, it is not clear at first why there is a strong attractive force between atoms which is able to hold the metal together against the mutual repulsion of the positive ion cores.

The secret lies in the wave nature of the electron and is illustrated diagrammatically in Fig. 1.10. An electron confined to an atomic stationary state is effectively confined to a small volume in space; this implies that the

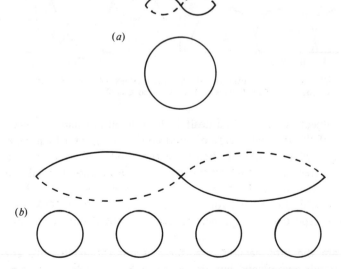

Figure 1.10 Relationship between momentum and wavelength of valence electrons for metallic bonding. (*a*) A single atom with small wavelength and large momentum; (*b*) a collection of atoms with large wavelength and small momentum.

electron must have a small wavelength and large momentum (and, therefore, high energy). An electron confined only by the boundaries of a solid, either as a stationary state or as a travelling wave, is free to have a much longer wavelength – and lower momentum – with a considerable saving of energy.†

We have already mentioned that the abundance of almost free electrons in metals leads to their well-known property of high electrical conductivity; this topic will be covered in detail in Part 2. Another property that follows from this is high thermal conductivity (to be covered in Part 3).

The metallic bond is non-directional in nature, so that the crystals form close-packed structures. By controlling these structures, one can tailor important mechanical properties of metals, such as strength and ductility. It is this control that leads to the importance of metals in modern technology. The materials science of metals will be introduced in Part 4.

Other types of bonding (covalent and van der Waals) may add to metallic bonding to give greater bond strength than the pure metallic bonding of sodium. This is the case for the transition metals, including those very important elements iron and copper.

In our slice of the periodic table, as we go from Na through Mg to Al, the bond strength and electrical conductivity increase. When we get to silicon, the bond strength is even higher, but suddenly there are very few electrons available for conduction; this arises from a change in the bond type. Silicon has covalent bonding.

1.10 The covalent bond

The covalent bond occurs when a valence electron from one atom forms a cooperative stationary state with a similar electron from a neighbouring atom. In Fig. 1.11, we illustrate the change in energy levels for silicon. In the free atom, the 3s state is full (two electrons of different magnetic moment signified by the direction of the arrows), whilst the 3p state has two electrons with four vacancies (four short of a closed shell). When the atoms are brought together, energy is saved by creation of new bonding states and all four electrons occupy these states.

The vacancies cannot be forgotten and, in fact, these form antibonding states. If we could give an electron the large amount of energy needed to occupy one of these states, it would be free of the bonds and so could conduct electricity. This is the first clue to the electrical properties of semiconductors and must be borne in mind when we come to Chap. 3, where we deal with electrical conductivity in more detail.

---

†The Heisenberg uncertainty principle also predicts energy saving. We know the position of the electron more precisely on an atom than in a solid. Our uncertainty of the electron momentum is, therefore, greater in the atom than in the solid. This implies, as the uncertainty is towards higher momentum, that the electron confined to an atom has more energy.

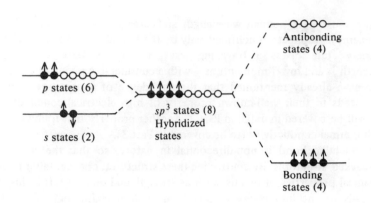

Figure 1.11 Change in valence electron energy levels in silicon when a covalent bond is formed. The hybridized states form at large separations and change to the bonding and antibonding states as the equilibrium separation is approached.

Each electron from a particular atom will usually form a combined state with a different neighbour; thus, each silicon atom interacts with four neighbours and, therefore, with a total of eight (shared) valence electrons. This gives a pseudo-closed shell configuration, which leads to the energy saving of the bonding states.

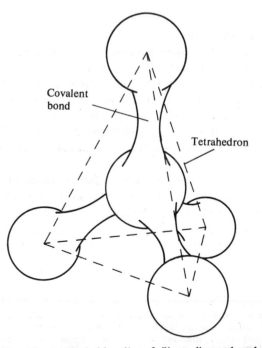

Figure 1.12 The tetrahedral bonding of silicon, diamond, and germanium.

This type of bond is strongly directional, because the number of bonds around an atom, and, therefore, the number of neighbours, is determined by the number of electrons available for bonding. In the case of an atom of silicon, the four neighbours will be arranged at the corners of a tetrahedron, as illustrated in Fig. 1.12.

Generally, an atom with n valence electrons can bond with $8n$ neighbours; thus, in silicon, a three-dimensional array of tetrahedral bonds are formed. A crystal of silicon can be regarded as a huge molecule, as the covalent bonding is continuous throughout the structure.

Phosphorus normally forms three covalent bonds and these can be accommodated in a planar hexagonal structure. Sulphur forms two covalent bonds and, therefore, has a one-dimensional chain structure. Chlorine can usually only form one bond, this creates the molecule Cl_2, which is saturated (all potential bonding states are occupied). The closed-shell elements like argon cannot form covalent bonds at all, each atom is a saturated molecule.

Now, although the bonding is weak, we know that all the elements in our example can exist in solid form — although we have to get to very low temperatures before argon solidifies. Some force must hold together the planes of phosphorus, the strings of sulphur, and the saturated molecules of chlorine and argon. This is the bonding force that always exists: the van der Waals or molecular bond.

1.11 The van der Waals bond

To explain this effect, let us again consider the electrons bound to an atom vibrating as standing waves in their different stationary states. If one could take a 'snapshot' of an atom, one would find, more often than not, that at any instant of time the negative electron charge is not distributed uniformly in space around the nucleus. Because of the vibration, the negative charge is biased towards one side. The atom is, therefore, an electric dipole. We can simplify this to a positive charge separated from a negative charge with dipole moment equal to charge x separation. This is illustrated in Fig. 1.13. The orientation of this dipole is constantly changing with the fluctuations of the electron vibration, but we can simplify this to a rotation. If the rotation of dipoles representing neighbouring atoms is in phase and, in the sense shown in Fig. 1.13(b), one can demonstrate that, as electrical forces obey the inverse square law (fall off rapidly as separation increases), positive poles are always nearer neighbouring negative poles than neighbouring positive poles; thus, there is a net electrostatic attractive force. This is the van der Waals bond. As atoms approach one another, the dipoles fall into phase, because there is a resultant lowering of the potential energy when this happens. It can be calculated, using the principles of electrostatics, that the potential energy saving (and, therefore, the bond strength) is inversely proportional to the sixth power of the separation between neighbouring dipoles.

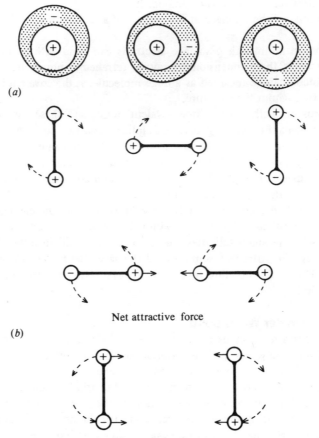

Figure 1.13 Atoms as rotating dipoles; the van der Waals bond. (*a*) Fluctuation of electron distribution and equivalent rotating dipole; (*b*) forces on neighbouring atoms.

As well as holding together the sheet, chain, and saturated-molecule solids, these forces can add to other forms of bonding, creating a much stronger bond in solids of greater use than solid argon.

For example, the van der Waals attraction in copper is ten times greater than that for krypton, the rare gas in the same row of the periodic table. This is because the additional metallic bonding results in an atomic separation in copper of 2.55 Å, whilst it is 4 Å for krypton, which has only van der Waals bonding. The inverse sixth-power law will give a much stronger van der Waals contribution to bond strength in the case of copper.

1.12   The ionic bond

This is the bond that can only be formed in compounds and it is the easiest to understand. In the ionic bond, a loosely-bound valence electron is completely removed from one atom, sodium in our example, and attached to the partner, chlorine in our example. In this way, both atoms acquire a closed-shell electron configuration, similar to neon for sodium and argon for chlorine. This process of electron transfer leaves the metal atom positively charged and the halogen atom negatively charged; the bonding arises from the electrostatic attraction between them.

The three processes and the energy balance involved in forming a molecule of sodium chloride are shown in Table 1.8.

In an ionic crystal, the ions arrange themselves so that the attractive force between ions of opposite charge is greater than the repulsive force between like ions. The crystal binding energy can be calculated almost exactly (unlike the case of covalent and metallic bonding). This is done by multiplying the electrostatic attraction between opposite ions with a constant (the Madelung constant) which depends on the geometry of the crystal structure.

The bonding of alkali halides is almost purely ionic (92 per cent for LiF and 94 per cent for NaCl), whereas the bonding of diamond, silicon, and germanium is 100 per cent covalent. Many compounds come somewhere in between these two extreme cases, with mixed covalent and ionic bonding. In other words, the valence electron states are shared, but are biased towards the atom with the greater electron affinity.† Examples of binary compounds of this type are:

SiC, 18 per cent ionic; GaAs, 32 per cent ionic; ZnO, 62 per cent ionic; AgCl, 86 per cent ionic.

A special example of mixed bonding is solid H_2O (ice). The electrons in the H–O bond form a shared (covalent) bonding state, but are mainly localized on the oxygen atom, leaving a net positive charge on the hydrogen atom with the oxygen atom being negatively charged. Opposite ends of neighbouring molecules are attracted, forming an ionic bond of the type usually called a hydrogen bond.

Table 1.8   Processes involved in the ionic bonding of sodium chloride

			Energy (eV)	
Process	Ingredients	Result	Used	Saved
Ionization	Na	$Na^+ + e^-$	5.1	
Electron capture	$Cl + e^-$	Cl^-		3.6
Bonding	$Na^+ + Cl^-$	NaCl		7.9

Net energy saved 6.4 eV

†The electron affinity is the energy given up when a neutral atom gains an electron to become a negative ion.

2. Crystal structures

2.1 Introduction

Although we started the book by consideration of the attraction between two atoms, we have already introduced the concept of solids composed of large assemblies of atoms in our models for the four types of bonding: the electron gas saving energy by occupying the whole volume of a metallic solid; the combined orbitals linking neighbour to neighbour, forming a network of atoms into a huge molecule as in the covalently-bonded solids; large assemblies of rotating dipoles held by molecular bonding; and the fine balance of electrostatic forces in ionic crystals.

We will not be concerned here with the complete set of geometrical arrangements that atoms bonded in these ways can take up, although this may be an interesting topic for further reading. We will simply describe the crystal structures of the more common and technically important solids and give a few interesting examples and engineering applications.

A regular crystal structure is the normal form of a solid. Minerals formed as the earth cooled from a liquid mixture. They are not a random arrangement of atoms, because the growing crystal will only accept an atom that fits well into the existing structure. Atoms bounce against each other about 10^{13} times every second; this trial and error process over millions of years leads to the unscrambling of the mixture and formation of large regular crystals in which each type of atom has its particular place. A single crystal is built up by a regular repetition of basic building blocks; hence, we can describe a crystal completely and know the position of every atom in the structure by describing the basic building block and the way it is repeated. But minerals have a complex structure, too complex to detail here, except for a brief description of the overall structure of some silicate minerals at the end of this chapter.

Man has extracted the elements from minerals for his own use and has formed synthetic compounds from them. These, generally, have a much simpler crystal structure than those of the naturally occurring minerals, simply because of the few types of atom involved. It is these structures with which we will be mainly concerned.

Before we can describe a crystal structure, we must learn something of the language used by crystallographers.

2.2 The language of crystallography

The terms used can be grouped into three groups of three.

A. Description of the overall structure (illustrated in Fig. 2.1).

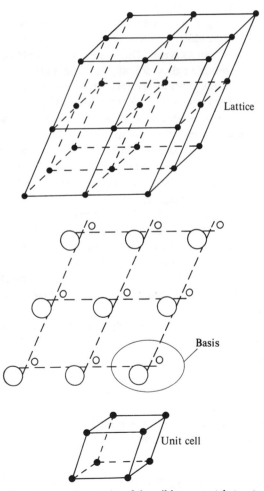

Figure 2.1   Three ways of describing a crystal structure.

1. *The lattice*   A regular arrangement of mathematical points in three dimensions described by the unit of repetition of the crystal. This unit, the distance between adjacent lattice points, is called the lattice constant. This is the basic unit of length in a crystal. If one could imagine a lattice for patterned wallpaper, this would be a two-dimensional arrangement of points, each point centred on one unit of the pattern.
2. *The basis*   The arrangement of atoms at each lattice point (if you like, the pattern on the wallpaper). The basis may be anything from a single atom, in the case of some elements, to a large organic molecule with 10^4 atoms, as in crystals of the proteins.
3. *The unit cell*   A three-dimensional cell constructed from adjacent lattice points that best describes the crystal structure and which will reproduce the whole structure by translation from one lattice point to another.

B. Conventions for naming planes, directions, and atom positions.

1. *Miller indices*   The Miller index convention is the convention for naming planes of atoms within a crystal. This system was founded in the eighteenth century, when minerologists found that the index numbers (defined below) of the outside faces of crystals (called the crystal habit) were small exact integers. They could deduce from this, although they did not have the modern X-ray or electron diffraction techniques, that crystals were made from a regular array of small building blocks.

First, in terms of the lattice constants, we must find the intercept of the plane we wish to describe with the three axes of the unit cell, *x, y, z*. The Miller index is then the reciprocal of these three intercepts reduced (usually) to the smallest integers, *h k l*.

In the examples given in Fig. 2.2, the plane cutting the axes at $x = 4$, $y = \frac{1}{2}$, and $z = 2$ is designated the (182) plane. In general, we will be describing much simpler planes. An example is the faces of a unit cell which are of the type {100}† as an intercept at infinity changes to 0 on taking reciprocals. For such an example, the planes parallel to the face diagonals are {110} and the planes containing two face diagonals from one corner are {111}.

2. *Crystal directions*   The set of smallest integers having the ratio of the components of the vector in the direction to be described. In the general case, the unit cell is *not* cubic, but most of the structures we will be describing *are* cubic and, in this *special case*, the normal to a plane is similar to the Miller index of that plane; thus, the normal to the (100) plane is the [100] direction.

3. *Atom coordinates*   These are determined by the distance travelled along each axis of the unit cell, in terms of the lattice spacing, to reach the atom

†Curly brackets { } mean 'of the type'; normal brackets ( ) mean the exact plane; square brackets [ ] signify a direction in the crystal.

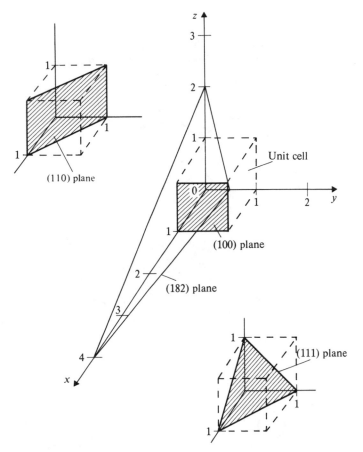

Figure 2.2   The Miller indices of atomic planes in a crystal.

in question. Thus, the origin, at one corner of the unit cell, is at 000 and an atom at the centre of the unit cell would have the coordinates $\frac{1}{2}$ $\frac{1}{2}$ $\frac{1}{2}$.

C. Three important quantities

1. *Coordination number*   The number of nearest neighbours to any atom in a crystal structure.
2. *Atomic radius*   Calculated assuming all atoms are spherical and touching their nearest neighbours.
3. *Packing density*   The fraction of the unit cell volume occupied by atoms, making the same assumptions as in C.2, above.

2.3   The close-packed structures

The majority of the elements are metals, and we know that metallic bonds are

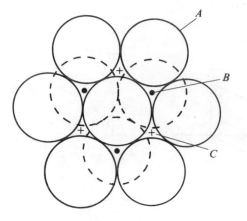

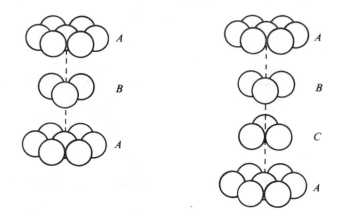

Figure 2.3   The two methods of stacking to achieve a close-packed structure.

non-directional; this is also the case for van der Waals bonds. To explain the crystal structures that arise in nature when the bonding is non-directional, we can use a simple model, where the atoms act like hard spheres and the energy is at a minimum when the maximum number of atoms are packed into a given volume†.

†Only close packing allows the maximum number of nearest neighbours separated by the equilibrium separation, and, therefore, the maximum bond-energy saving.

A simple way to demonstrate this model is by using a box containing a large number of steel ball bearings. These may be jumbled-up at first; however, if the box is shaken, to simulate high temperatures, the balls rearrange themselves into a regular pattern, forming a close-packed structure. This structure is illustrated in Fig. 2.3.

The atoms form close-packed planes where each atom in the plane is touching a hexagon of six similar atoms. If we now think of the way these planes can fit on top of one another, we can start with a close-packed plane (A), and will find that the next plane (B) will sit so that each atom is nestling between, and is, therefore, touching, three atoms of plane (A). Where will the next plane go? If we look carefully, we find that we now have a choice of two alternative positions which are equally attractive in our hard-sphere model. In nature, because of other effects, one of these alternatives is chosen exclusively throughout the crystal. The two alternatives lead to the different crystal structures described below.

2.4 The hexagonal close-packed structure

The first alternative is that the third plane is placed directly over the first plane (A) and we have a crystal with $ABABAB\ldots$ packing. This structure is the hexagonal close-packed (h.c.p.) structure. As each atom is touching six atoms in its own plane, three in the plane above and three in the plane below, the coordination number of a close-packed structure is 12. A unit cell of the h.c.p. structure is shown in Fig. 2.4(a). The packing planes (A and B) are of the {100} type. The unit cell shown is a prism of hexagonal cross-section and is complicated by the fact that it has three axes in the base plane (100) at 120° to

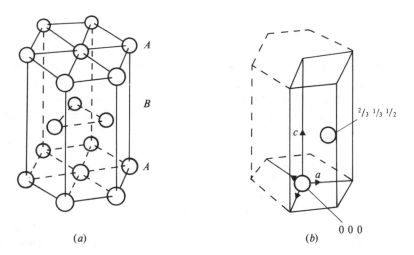

(a)

(b)

Figure 2.4 The close-packed hexagonal structure. (a) The conventional unit cell; (b) the primitive unit cell.

Table 2.1 The crystal structures of the elements

Crystal structure	Group or type	Elements
Hexagonal close-packed	Groups 2a and 2b	Be, Mg, Ca, Sr, Zn, Cd
	Transition metals	Sc, Ti, Y, Zr, Hf, Tc, Ru, Re, Os, Co
	Also	He (Group 0) and 9 Lanthenides
Face-centred cubic	Group 1b	Cu, Ag, Au
	Group 8	(Fe, Co) †, Ni, Rh, Pd, Ir, Pt
	Rare gases	Ne, A, Kr, Xe, Rn
	Also	Al (Group 3a) and Pb (Group 4a)
Body-centred cubic	Alkali metals	Li, Na, K, Rb, Cs, Fr.
	Transition metals	Ti, V, Cr, Y, Zr, Nb, Mo, Hf, Ta, W.
	(Groups 3b, 4b, and 5b)	
	Also	Sr. Ba (Group 2a), Fe (Group 8) and F (Group 7a)
Diamond	Group 4a	C, Si, Ge, Sn
Other structures	The non-metals	N, P, As, Sb, Bi (Group 5a) O, S, Se, Te, Po (Group 6a) Cl, Br, I, At (Group 7a)
	Also	H (1a), Ra (2a), B, Ga, In (3a) Hg (2b), Mn (7b), two lanthenides and five heaviest actinides (where structure known).

† Above room temperature

each other, conventionally called the a axes, and one axis normal to the base plane, called the c axis. The lattice spacing is greater in the c direction than in the a direction because of the B plane sandwiched between the two A planes.

The unit cell shown is not the cell of minimum volume that would reproduce the crystal structure. The smallest unit cell (or primitive cell) is illustrated in Fig. 2.4(b) and, in terms of this cell, the lattice is *simple* hexagonal with a basis (repeated at each lattice point) of two identical atoms; one at the origin 0 0 0 (representing the A-plane atoms) and one at $\frac{2}{3} \frac{1}{3} \frac{1}{2}$ (representing the B-plane atoms). Conventionally, the larger cell, as shown in Fig. 2.4(a), is taken as the unit cell, as this gives a clearer idea of the structure. (The same situation occurs in the case of the two cubic structures described later.) The crystal structures of the elements are listed in Table 2.1. One can pick out some useful elements with an h.c.p. structure. For example, zinc and cadmium are used for coating steel items to protect them from corrosion. Magnesium is an important component of alloys with a high strength to weight ratio.

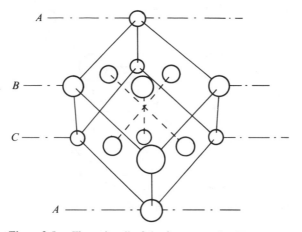

Figure 2.5 The unit cell of the face-centred cubic structure.

2.5 The face-centred cubic structure

The second alternative in the packing of close-packed planes is that the third plane is placed in position *C* (Fig. 2.3). This is a position not used for planes *A* and *B*. Once more, the coordination number is 12. The fourth plane is then placed in position *A* and we have a crystal packing as *ABCABCABC* . . . and so on. This is called the face-centred cubic (f.c.c.) structure, as the unit cell (Fig. 2.5) is a cube with atoms at the corners of the cube and at the centre of the faces.

It is not easy to see how the unit cell lies in relation to the close-packed planes *A*, *B*, and *C*, because, in this case, the close-packed planes are of the {111} type. With reference to Fig. 2.5, one can see that, if one corner atom is in the *A* plane, the three corner atoms and the three atoms at the centre of faces closest to this atom are in the *B* plane, the next three corner atoms and the three remaining face-centred atoms are in the *C* plane, and the corner atom diagonally opposite to the first atom is again in the *A* plane.

Many useful metals have the f.c.c. structure and, with the exception of aluminium and lead, they come from Group 8 and the neighbouring Group 1b of the periodic table. For the electrical engineer, there are the conductors: copper, aluminium, silver, and gold. For the mechanical engineer, aluminium is used extensively in engine castings and alloying additives such as nickel are used in making steel. For the civil engineer, there are the corrosion-resistant metals, i.e., lead, copper, and copper–nickel alloys, for roofing and plumbing.

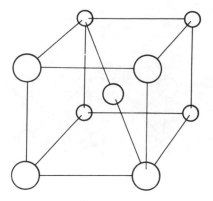

Figure 2.6   The unit cell of the body-centred cubic structure.

2.6   The body-centred cubic structure

Our simple model for close-packed structures arising from non-directional bonding breaks down for many elements, especially the alkali metals and the early groups in the transition series. These do not crystallize in close-packed structures. Instead, they have the body-centred cubic (b.c.c.) structure. A unit cell of this structure is illustrated in Fig. 2.6. The unit cell is a cube with an atom at each corner and one at the body centre.There are eight nearest neighbours to each atom rather than 12 as in the close-packed structures.

Although this structure is not close-packed, we will show later that it is *almost* close-packed.

Iron, towering in importance as a basis of alloys useful in all aspects of engineering, has the b.c.c. structure at room temperature. The metals used for service at very high temperatures in the aerospace industries, i.e., the 'refractory' metals molybdenum, tantalum, and tungsten, also have this structure.

2.7   Second-neighbour effects

The breakdown of our simple model for structures arising from non-directional bonding occurs as a result of the fact that the energy-minimizing process is not dependent only on the position of nearest neighbours. The interatomic forces extend beyond this and the so-called hard-sphere model of an atom is not completely appropriate.

The position of the next-nearest neighbours and even more distant neighbours influences the structure. This is why the b.c.c. structure forms and is also why, generally, in nature, we do not get mixed close-packed stacking.† In nature, *ABABAB. . .* or *ABCABCABCABC. . .* prevails throughout the crystal.

Many elements have several alternative structures. The energy balance may differ, depending on the kinetic energy of the atoms (temperature). For example, cobalt transforms from the h.c.p. form to f.c.c. when the temperature is raised to 130 °C. Iron is b.c.c. below 910 °C, f.c.c. between 910 °C and 1400 °C, and returns to b.c.c. above 1400 °C.

This can be taken as evidence that the energy differences between the two close-packed structures h.c.p. and f.c.c. and the almost close-packed b.c.c. structure are small.

2.8 The diamond structure

When we have covalent bonding, crystal structures are formed that are very different to the close-packed structures. The number of valence electrons determines the number of bonds and, therefore, the coordination (number of nearest neighbours). Because of this, the coordination number and packing density of covalently-bonded solids is, generally, low. The crystal structure is built up and energy is minimized by satisfying the geometry of the bonding rather than by close-packing of atoms. In the case of the Group 5, 6, and 7 elements given as having 'other structures' in Table 2.1, we have mixed bonding: strong covalent bonding within the sheet, chain, or molecule and weak van der Waals bonding holding these units together to form a three-dimensional crystal of rather complex symmetry. Examples of structures of this type will be given at the end of this chapter when some silicate minerals are described.

The extreme case of covalent bonding is that of the Group 4a elements, C, Si, Ge, Sn. Here, four covalent bonds are formed, creating a tetrahedron of nearest neighbours. As the bonds continue in three dimensions, no other form of bonding is needed to create a crystal. The crystal can be thought of as one giant molecule.

The crystal structure of the Group 4a elements is called the diamond structure, and a unit cell of this structure is illustrated in Fig. 2.7. This can be analysed as a face-centred cubic lattice with a basis of two atoms, one at the lattice point, i.e., at 0 0 0, and one at $\frac{1}{4}$ $\frac{1}{4}$ $\frac{1}{4}$. When this basis is repeated at each of the lattice points, we build up the unit cell. This contains four tetrahedra, two in opposite corners of the bottom half and two in the alternative opposite corners in the top half. The four corner atoms, apparently not bonded, have all four bonds in neighbouring unit cells. Of course, as in the case of the other structures we have discussed, it does not matter which atom we choose as the origin of the unit cell. As each atom is identical and is in an identical

---

†So-called *stacking faults* do occur, but not often enough to warrant description here. These will be briefly discussed in Chap. 7.

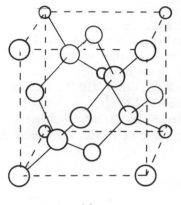

(a)

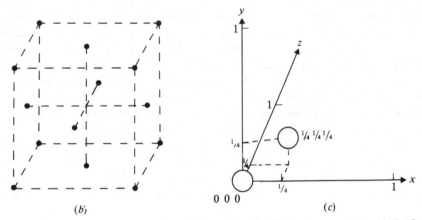

(b)   (c)

Figure 2.7   The diamond structure. (a) The unit cell; (b) the face-centred cubic lattice; (c) the basis, 0 0 0 and $\frac{1}{4}\frac{1}{4}\frac{1}{4}$.

environment, we would always obtain the arrangement of atoms shown in Fig. 2.7.

2.9   Comparison of the cubic structures

It is interesting to compare the coordination number and packing density of the three cubic structures already described: f.c.c. (close-packed), b.c.c. (almost close-packed), and diamond (directional bonding). This is done in Table 2.2, which includes the simple cubic structure with a unit cell of a cube with an atom at each corner. This is a structure that rarely occurs in nature (of the elements, only oxygen, phosphorus, and manganese have a simple cubic lattice), but it will serve to illustrate how the quantities in Table 2.2 are calculated.

Table 2.2 Properties of the cubic crystal structures

Quantity	f.c.c.	b.c.c.	Simple cubic	Diamond
Coordination number	12	8	6	4
Atoms per unit cell	4	2	1	8
Atomic radius †	$0.35a$	$0.43a$	$0.5a$	$0.22a$
Packing density (per cent)	75	67	52	36

† The atomic radius is given in terms of the lattice constant a.

Consider a simple cubic structure, as shown in Fig. 2.8. The coordination number is six. There are eight corners to a cube and each atom is shared by eight neighbouring cubes; therefore, there is effectively $8 \times \frac{1}{8} = 1$ atom per unit cell. In Fig. 2.9, we depict the atomic surface that intersects the unit cell boundaries with the atoms drawn more realistically to size (they were drawn undersized in earlier figures to enable the structure to be seen more clearly). We can see that the atomic radius for a simple cubic structure is just half the lattice constant a. Thus, the volume of the unit cell occupied by atoms is $\frac{4}{3}\pi(0.5a)^3$, whereas the volume of the unit cell is a^3. The ratio of these two numbers gives a packing density of 52 per cent.

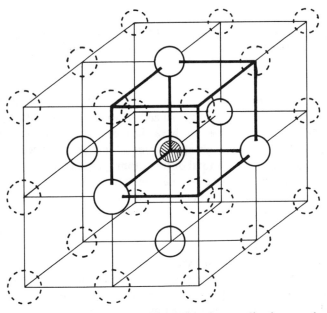

Figure 2.8 The simple cubic structure, illustrating the coordination number and the number of atoms per unit cell.

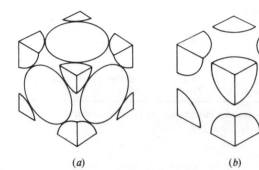

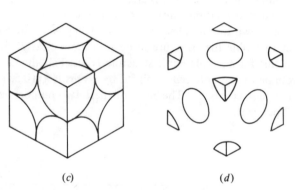

Figure 2.9 The intersection of atoms drawn to scale with the cell boundaries. (*a*) Face-centred cubic; (*b*) body-centred cubic; (*c*) simple cubic; (*d*) diamond.

It can be easily deduced from Figs. 2.5 and 2.9 that the atomic radius for f.c.c. is one quarter of the face diagonal and, as well as the one atom per unit cell representing the corner atoms, we have six face-centred atoms each shared by two neighbouring unit cells, giving us three atoms per unit cell from the face centres. We have, therefore, a total of four atoms per unit cell.

The atomic radius for b.c.c. is one quarter the body diagonal and for the diamond structure it is one eighth of the body diagonal.

It is interesting to see that the packing density increases as the coordination number increases.

2.10 Crystal structures of compounds

Just as there are a very large number of possible compounds, there are a very large number of possible crystal structures, although, from symmetry considera-itons, these can only be accommodated on one of 14 lattices. (We have described the three most common, f.c.c., b.c.c. and h.c.p.; simple cubic is also one of the 14.)

We will describe some important crystal structures of simple compounds where the basis is a molecule of the compound and the lattice is one of the types already described. Each structure is usually typical of a family of compounds of similar bonding.

We will describe the crystal structure of three AB-type compounds: NaCl, CsCl (ionic bonding), and ZnS (covalent bonding). Examples of two AB_2 compounds are also given: SiO_2 (in the form of β crystabolite) and CaF_2 (fluorite).

2.11   Two ionic AB compounds

The two purely ionic compounds taken as examples have quite different structures. This seems strange at first, because we know that the valency and bonding is identical in both NaCl and CsCl.

The only difference is that the sodium ion has the same size as an atom of the inert gas neon, smaller than the chlorine ion, whilst the caesium ion has the size of the heavy rare gas atom, xenon, much larger than the chlorine ion. The difference in ion size means that the crystal structure arising from the necessity

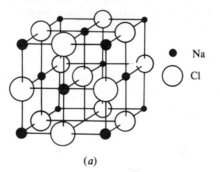

$\bullet$ Na

$\bigcirc$ Cl

(*a*)

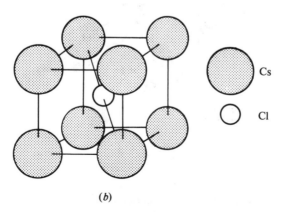

Cs

Cl

(*b*)

Figure 2.10   Unit cells of two ionic AB compounds. (*a*) NaCl; (*b*) CsCl.

to balance the electrostatic forces and to achieve electrical neutrality is different in the two cases.

The two structures are illustrated in Figs. 2.10(*a*) and (*b*). In the case of NaCl, the chlorine ions are almost touching in a close-packed structure, with the sodium ions fitted into the void left between six neighbouring chlorine ions. The lattice is face-centred cubic, with a basis of Na at the origin 0 0 0 and Cl at the body centre $\frac{1}{2}\frac{1}{2}\frac{1}{2}$. This results in two f.c.c. structures, one of sodium and one of chlorine, displaced from one another by half the unit cell body diagonal.

The caesium ion is large and can, therefore, be surrounded by eight chlorine ions. This results in a *simple* cubic lattice and a basis of Cs at 0 0 0 and Cl at $\frac{1}{2}\frac{1}{2}\frac{1}{2}$. Thus, NaCl and CsCl have similar bases, but the lattice is f.c.c. in the former case and *simple* cubic in the latter.

Compounds with the NaCl structure include KCl, PbS, AgBr (important in photography), and the most refractory of the ceramic oxides, MgO (magnesia).

Important examples of compounds with the CsCl structure are NH_4Cl (the ammonium ion NH_4^+ acting as one unit), and intermetallic compounds such as CuZn (β-brass), AlNi, and the hard alloy BeCu used in making springs and parts with good strength and high electrical conductivity. It is interesting to note how a structure analogous to the b.c.c. structure common amongst metals is copied by intermetallic compounds

2.12   The zinc sulphide structure

The structure of ZnS (or the zinc blende structure) is shown in Fig. 2.11. This is very similar to the diamond structure, with the same lattice (f.c.c.) but with a basis of Zn at 0 0 0 and S at $\frac{1}{4}\frac{1}{4}\frac{1}{4}$.

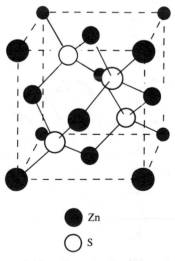

Zn

S

Figure 2.11   The unit cell of zinc sulphide.

This structure often occurs where the average valency per atom is four, allowing the tetragonal bonding to occur. An example from Group 4a is SiC, the very hard material used for grinding and cutting tools. Important examples of semiconductors with this structure, formed by compounding Group 3a and Group 5a elements (3—5 compounds), are GaP and GaAs. These semiconductors can change electrical signals into light and produce very high-frequency electromagnetic waves. This structure also occurs for 2—6 compounds, ZnS of course, but also two more of the family of compound semiconductors: CdS and CdTe. It must be remembered, from the examples given in Chap. 1, that the bonding is not entirely covalent for the compound semiconductors. There is always an appreciable amount of ionic bonding as well.

2.13 Two AB$_2$ compounds

The most common compounds on the Earth's crust are based on the SiO$_2$ molecule. This is because silica is one of the lightest solids and so floated to the top surface of the molten embryo earth. The building block of the silicates is a tetrahedral arrangement of four relatively large oxygen atoms with a relatively small silicon atom nestling in the void between them (Fig. 2.12(a)). There are six crystal structures based on this building block. One of these is quartz, which has a hexagonal lattice and is used as the vibrating heart of resonators for electronic clocks and watches because it produces a pulse of exceedingly high regularity when the crystal is made to vibrate. Most of the six have hexagonal symmetry

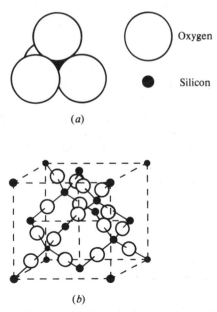

Figure 2.12   Silicates. (a) The SiO$_4^{4-}$ tetrahedron; (b) the unit cell of β-crystabolite (SiO$_2$).

48

and are too complex to describe here, but one has an f.c.c. lattice with silicon arranged on a diamond structure and oxygen arranged halfway along the tetrahedral bonds, i.e., at $\frac{1}{8}$ $\frac{1}{8}$ $\frac{1}{8}$ positions, this is called β-crystabolite and is illustrated in Fig. 2.12(*b*). Of course, the unit cell has to be much bigger than that for pure silicon in order to accommodate the oxygen.

Another AB_2 compound is fluorite, CaF_2. This has an f.c.c. lattice of calcium, with fluorine occupying all eight $\frac{1}{4}$ $\frac{1}{4}$ $\frac{1}{4}$ tetrahedral positions. Note that this differs from the ZnS structure because, in the latter case, only *four* of the tetrahedral sites are occupied.

2.14 Cases of crystal use

(a) Electrical engineering

The electrical engineer makes good use of crystals for their unique properties. A prime example of this is that of the semiconductors. It is only through the development of super-perfect crystals, especially of silicon, that the age of the transistor and the integrated circuit has arrived. Only when we have a pure, perfect crystal can we alter the electrical properties controllably to create all the circuit functions required by the electronic engineer. We have said that a crystal of silicon is like a giant molecule, but this has to stop at a surface, leaving a mass of unsatisfied bonding electrons. It is the satisfaction of these 'dangling bonds' by growth of a pure oxide that has allowed the newest active device, the metal-oxide-semiconductor transistor (or MOST) to be developed. Two micrographs of a silicon integrated circuit are shown in Fig. 2.13.

At the other end of the spectrum, there is a requirement in electrical motors and generators for brushes of high electrical conductivity but good lubricating properties (slipperiness). Graphite, the common form of carbon, is used and the secret of the ideal blend of properties lies in the crystal structure. In graphite, the carbon forms three covalent bonds with neighbours, and, therefore, the crystal structure is one of planar hexagonal arrays with covalently-bonded carbon atoms, as shown in Fig. 2.14. The planes are held together by weak van der Waals bonds and can, therefore, easily slip over each other. This is the secret of the good lubricating properties of graphite. The fourth unsatisfied electron is in an antibonding state, and is free to take part in electrical conduction.

Yet another case where the crystal structure is the key to understanding unique electrical properties is that of the ferroelectric materials such as barium titinate, $BaTiO_3$. A unit cell of this structure is shown in Fig. 2.15. This is the perovskite structure ($CaTiO_3$), with the calcium replaced by barium. The unit cell has barium at the cube corners, oxygen at the face centres, and titanium at the body centre. This material is called a ferroelectric because the material polarizes spontaneously by displacement of the central titanium atom — a displacement which occurs in the same sense for all the titanium atoms. This is analogous to the spontaneous alignment of magnetic moments in a ferromagnet.

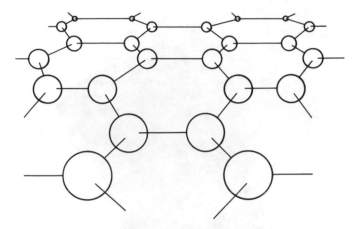

Figure 2.14   The planar structure of graphite.

As the capacitance of a dielectric is related to the ease with which electric dipoles can be created within the dielectric, and large dipoles can be easily created in the ferroelectrics, they are very useful in making large capacitors.

(b) Civil engineering
Civil engineers have traditionally used the materials that commonly occur in the Earth's crust because of their large quantity and low cost. These are the minerals which are mainly based on the SiO_4^{4-} tetrahedron (the 4– signifies a surplus bonding electron on each oxygen ready to form bonds to the next unit of the structure), which is shown in Fig. 2.12(a).

Three-dimensional structures like quartz are important components of hard stone and gravel used for building materials.

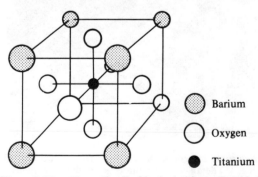

○ Barium

○ Oxygen

● Titanium

Figure 2.15   The unit cell of barium titinate ($BaTiO_3$).

Asbestos, that useful but unhealthy insulating material, is constructed of double or single chains of SiO_4^{4-} tetrahedra, bonded weakly by planes of hydroxyl ions and metal ions like aluminium and potassium. Mica crystals (Fig. 2.16) are readily parted (cleaved), as sheets along the weak planes. Four layers of silicate sheets, the bread in the sandwich, separated by three layers of jam (OH⁻,

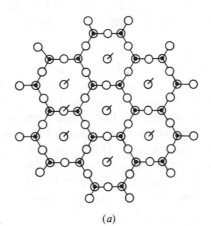

(a)

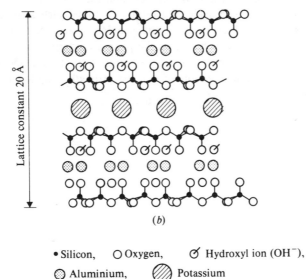

(b)

• Silicon,   ○ Oxygen,   ⌀ Hydroxyl ion (OH⁻),

◉ Aluminium,   ◫ Potassium

Lattice constant 20 Å

Figure 2.16   The structure of mica. (a) Plan view of a silicate sheet; (b) side view showing position of OH⁻, Al⁴⁺, and K⁺ ions.

Al^{4+}, and K^+ ions) are needed to make up one unit cell, so the lattice constant for mica is large. Two other minerals with structures only slightly different to this are talc and kaolinite. Talc has good lubricating properties for the same reasons as graphite, the planes readily slip over one another. Kaolinite is more commonly known as clay, the basic component of bricks and pottery. Water molecules lie between the sheets, held by van der Waals forces. These allow the silicate sheets to slip across one another in a way that gives wet clay its characteristic plasticity. If too much water is added, water—water bonds form, plasticity is lost, and a soupy liquid called 'slip' is formed.

(c) Mechanical engineering

Most metals in service are made of many small crystals, they are polycrystalline. Each crystal is called a grain and an annealed metal is usually composed of a collection of grains of all different orientations. The strength of a single crystal, measured by the modulus of elasticity (how much stress or loading is needed to stretch it a given amount), varies with orientation; for example, the modulus of elasticity for the b.c.c. structure in the [111] direction is more than twice that in the [100] direction. This is because the {111} planes are the close-packed planes in the b.c.c. structure and, by pulling in the [111] direction, which is normal to the (111) plane, we are trying to pull them apart.

In a normal polycrystalline metal, these effects even out, but occasionally this crystalline effect can be put to good use.

An example of where this is done comes from the jet turbine engine. The turbine blades, especially those in the combustion section, have to work at very high temperatures. These are made of a refractory metal such as tungsten. The original blades were polycrystalline, but problems were encountered with uncontrolled grain growth and failure at the grain boundaries. This problem has been overcome by modifying the casting technique so that the blades are grown as single crystals with the direction of greatest strength oriented to meet the maximum stress.

Part 2

Electrons in Solids

'It may be a sound —
A tone of music — summer's eve — or spring
A flower — the wind — the ocean — which shall wound,
Striking the electric chain where with we are darkly bound.'

('Childe Harold' *xxiii* by Lord Byron)

'While this magnetic
Peripatetic
Lover, he lived to learn,
By no endeavour
Can magnet ever
Attract a Silver Churn!'

('Patience' by W. S. Gilbert)

3. Electrical conduction

3.1 Introduction
In this chapter, two approaches, or models, are used to explain the concepts behind the vastly different electrical properties of conductors (metals), insulators, and semiconductors.

These models come together at the end of the chapter and are used to explain such diverse phenomena as electrical resistance, the doping of semiconductors to produce transistors, and conversion of electrical signals into light.

In the first model, we treat the electron as a stationary state bound to an atom. We give examples of types of the stationary state, describe the energy levels in a single-electron atom, and explain how these are modified in a many-electron atom. We then describe how these stationary states are modified when we bring the atoms together to form a solid. This is called the *tight-binding* approach because the electrons start tightly bound to a single atom.

In the second model, the valence electrons are treated as a free gas, confined only by the walls of the solid. The properties of this gas are determined ignoring the atomic cores (ions) left by the valence electrons. Then, the effect of the atomic cores on the allowed energies of the *almost* free electrons is described. This is the *loose-binding* approach.

3.2 Stationary states of electrons bound to atoms
In Chap. 1, we introduced quantum numbers by studying the modes of vibration of a string bound at the ends: the violin string. We found that each allowed vibration had a well-defined energy which was proportional to the square of the quantum number (the order of the resonance). We stated that an electron in an atom was also a bound resonant system called a stationary state, but, as the vibration was in three dimensions, and rather complex, the relationship between energy and quantum number was different to that for the violin string. As well as this, there are several types of stationary state for bound electrons, namely the s-, p-, d-, and f-type 'orbitals'.

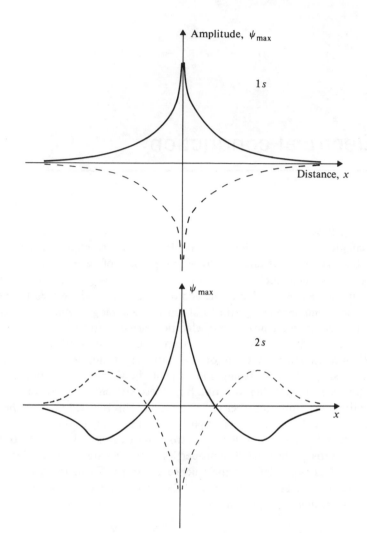

Figure 3.1 Maximum amplitude (ψ_{max}) versus distance (x) for the electron stationary states 1s and 2s.

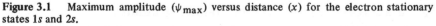

We will now describe, as best we can,† some of the simpler stationary states. The two simplest states are the 1s and 2s states; these are illustrated in Fig. 3.1, where the two positions of maximum amplitude of the electron wave (ψ_{max})

---

†We really need *four* dimensions to picture a stationary state – the three dimensions of space and one of amplitude or intensity of vibration, but on paper we have only two and can perhaps hint at a third.

are shown. The *s* states are spherically symmetric, so that, by examining the vibration in just one dimension, we can visualize the vibration in three dimensions. The one-dimensional representations of the 1*s* and 2*s* states are similar to the two orders of vibration *n* = 1 and *n* = 3 of the violin string (Fig. 1.6). One must remember, though, that the 1*s* state in three dimensions is a sphere with maximum vibration at the centre, diminishing towards the boundary. The 2*s* state is, again, a sphere, but with a maximum vibration at the centre and another maximum on a spherical surface near the boundary, the two maxima being separated by a sphere of zero vibration. As the confinement of the electron in space is more subtle than the clamping of a string, the cut-off at the boundaries is not sharp, but more like an exponential decay to smaller and smaller amplitudes. This explains why atoms can attract one another at quite large separations; the electron stationary states extend well beyond the 'edge' of an atom, although they have a very small amplitude once one is far away from the atom.

One can also represent a stationary state by plotting the square of the maximum amplitude† of the stationary state against distance. This plot (which is never negative) is called the probability density, as it can be thought to represent the probability of finding an electron a certain distance from the nucleus. As we know, the *whole* stationary state *is* the electron, so, alternatively, we can consider the plot to represent simply the degree of vibration of the standing wave.

The maximum amplitude and probability density in one and two dimensions are shown in Fig. 3.2 for 1*s*, 2*s*, 2*p*, 3*p*, and 3*d* states. The point of maximum vibration in the two-dimensional representation is indicated by the letter M.

The 2*p* state is not spherically symmetric and can be thought of as two ellipsoids of vibration separated by a node which is a flat surface going through the centre of the atom. The lack of symmetry of the 2*p* stationary state is the reason for this state exhibiting an angular momentum (and, therefore, a magnetic moment) and there are three possible orientations of this state (for reasons which are not obvious here); hence, we have three possible values of the resultant magnetic moment in any particular direction.

The higher-order states tend to have a version of the appropriate lower-order state at the centre, with a related area of high probability density around this. Thus, the 2*s* state has a '1*s*-like' region at the centre, with a spherically symmetric vibration around this. The 3*p* state has a '2*p*-like' region at the centre, with two more regions of oscillation outside this.

The areas of high probability density for the lowest-order *d* state, 3*d*, are rather like two eggs and two sausages arranged in a cross with the sharp ends pointed at the nucleus. Like the *p* states, the *d* states have a node at the nucleus,

---

†To be correct, we actually multiply the wave function ψ by its complex conjugate ψ^*, which gives us a real number (the wave function includes complex numbers).

58

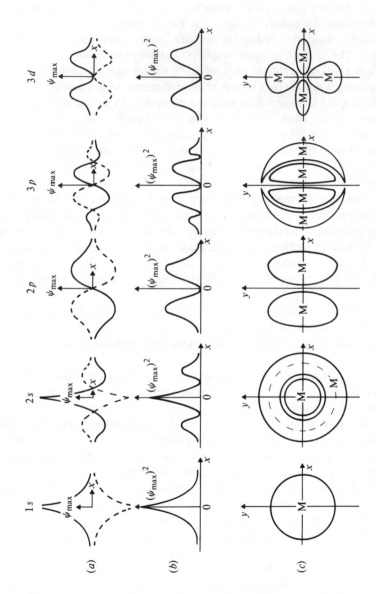

Figure 3.2 The electron stationary states 1s, 2s, 2p, 3p, and 3d. (a) The maximum amplitude (ψ_{max}) versus distance (x); (b) the probability density (ψ_{max})^2 versus (x); (c) contours of constant probability density in the xy plane, position of maximum probability marked M.

exhibit angular momentum (five possible orientations in this case), and the higher-order states (not shown) have a central region similar to the lower-order state surrounded by related regions.

3.3 Energy levels of electrons bound to atoms

(a) The single-electron atom

The energy level scheme for a one-electron atom (hydrogen) is shown in Fig. 3.3(a). The first thing to observe is that, as the energy increases, the levels get closer together, as predicted by the Bohr model discussed in Sec. 1.5.

The more complex a vibration becomes, the smaller the energy difference between different states. Alternatively, we can say that at higher energies we have more states per unit energy, i.e., a higher density of states. This idea will be used again when we come to the loose-binding model. In fact, for large values of the principal quantum number n, the levels are so close together that almost any value of energy is allowed. When this occurs, we say that we have a continuum of energy, and an electron with this energy would not be bound to the atom.

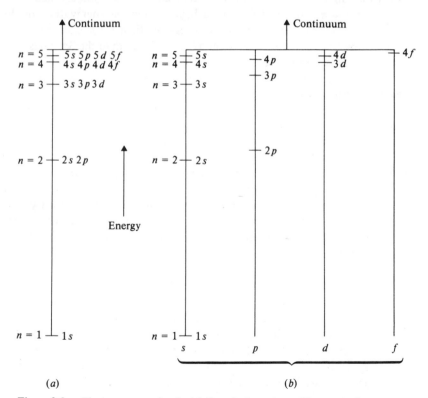

(a)   (b)

Figure 3.3   Electron energy levels. (a) One-electron atom; (b) a many-electron atom.

If the electron which normally occupies the $n = 1$ 'ground' state were given some energy, it would change oscillation to that of the appropriate higher state (called an excited state). If it were given enough energy to reach the continuum, it would be free of the atom altogether and become a travelling wave.

The second thing to note from Fig. 3.3(a) is that only the *principal* quantum number determines the electron energy for a one-electron atom.

(b) The many-electron atom

The situation is different for an atom containing many electrons, because the electrons interact with each other and this affects the energy of the stationary states. As the s-type states have a maximum probability at the nucleus, they will benefit from the full attraction of the nucleus with a total charge equal to the total number of electrons. This will cause the s states to draw closer to the nucleus and give them a lower energy than in the case of the single-electron atom, because they sit in a deeper potential well. The p, d, and f states have nodes at the nucleus, so the electrons in these states only see the nucleus through the intervening s states, which veil some of the attractive force of the nucleus. Thus, the p, d, and f states experience an attractive force which is less than that for the s states and so have a higher potential energy than in the one-electron atom. The effect for d states is greater than for p states, which is why, for example, we have the $4s$ states at a lower energy than the $3d$ states. The increase in energy is even greater for the f states.

The energy-level scheme for the lower energy levels of a many-electron atom is illustrated in Fig. 3.3(b).

It is still the case that more than one electron can occupy each level (two for s, six for p, 10 for d, 14 for f), but the Pauli exclusion principle dictates that each electron must have a unique set of four quantum numbers. However, two of these (the ones related to magnetic moments) do not denote a difference in energy in the absence of a magnetic field; they are termed *degenerate*.

If a magnetic field were applied, the levels would split further, giving a separate level for each value of electron magnetic moment and stationary-state (misleadingly called 'orbital') magnetic moment. In this case, the degeneracy will have been removed and a difference in quantum number truly represents a difference in energy.

3.4 Combination of stationary states in solid

We will now describe what happens when atoms bond together to form a molecule or a solid. The formation of a bond means that the valence electrons interact with one another. This results in the formation of combined stationary states, which, in turn, means that the solid as a whole can be considered as having energy levels; however, the energy-level diagram looks very different to that for single atoms where valence electrons are concerned.

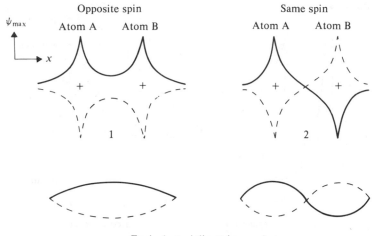

Equivalent violin-string modes

Figure 3.4   Combination of two 1s states from neighbouring hydrogen atoms.

(a) Two hydrogen atoms

We will take the simplest stationary state, the 1s ground state of the hydrogen atom. What happens when two atoms bond together to form a molecule?

Because of electron 'spin', we can have one of two possible combined states; these are illustrated in Fig. 3.4. Spin is a term used to describe the magnetic moment† of the electrons. The quantum number can have two values ('spin up' and 'spin down') and this is the reason why two electrons can occupy the s states and still obey the Pauli exclusion principle.

In our illustration, the 1s states from neighbouring atoms are combining. For state 1, where the electrons have opposite spin, the exclusion principle is obeyed and energy is saved by bringing the atoms together, i.e., a bond is formed. The variation of energy with interatomic separation is shown in Fig. 3.5 and is similar to that shown schematically in Fig. 1.8, when interatomic bonding was first discussed.

State 2 is an antibonding state. Since the electrons have the same spin, the exclusion principle is disobeyed, and there is a repulsive force between the atoms which increases as interatomic separation decreases. This is shown as the upper curve in Fig. 3.5.

It is interesting to note, from Fig. 3.4, that there is an appreciable amplitude between the atoms in the bonding state. Indeed, the combined oscillation is

†Not till Chap. 4 will magnetic moment be defined. For now, you are asked to accept this as a property of an electron which is denoted by the spin quantum number.

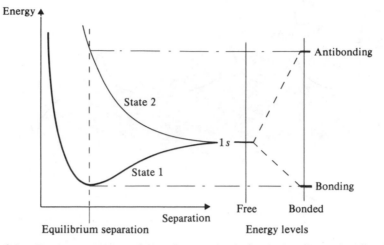

Figure 3.5 Energy versus separation and energy levels for the bonding and antibonding combined 1s states of two hydrogen atoms.

rather like the lowest energy state for the violin string ($n = 1$). The antibonding state has a node between the atoms and the combined oscillation is like the higher-order vibration of the violin string ($n = 2$).

(b) Six hydrogen atoms

Now, let us increase the complexity of our molecule and see what combined states we can obtain for a 'solid' containing six atoms of hydrogen.

The exclusion principle must be obeyed for the combined states just as for atomic states. Therefore each state combining all six atoms can contain only two electrons, one of spin up and one of spin down. Each atom has one s electron and one s electron vacancy in the 1s shell. For six atoms, we have six electrons and six vacancies to fit into stationary states. This implies that we need six combined stationary states in our six-atom solid.

A one-dimensional picture of these states, at maximum amplitude in one direction, is shown in Fig. 3.6 for six atoms in a straight line. This illustration has been constructed by modifying the neighbouring 1s states and making them conform to an 'envelope' corresponding to the violin-string modes for values of n from 1 to 6. Thus, even at this fairly low level of complexity, we see that, for the lowest energy states, electron wavelength is increasing as the number of atoms increases; the states look more and more like violin-string states.

The variation of energy with separation for these six states is shown in Fig. 3.7. The six electrons can be seen to occupy the three lower-energy bonding states in the ground state of the solid. However, the states are now quite close together, so that not much energy is needed to promote an electron from the

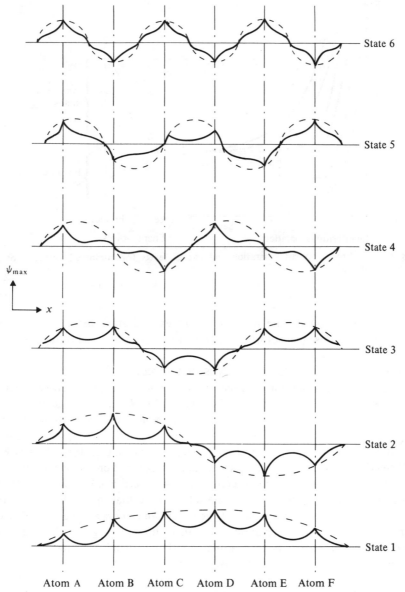

State 6

State 5

State 4

ψ_{max}

x

State 3

State 2

State 1

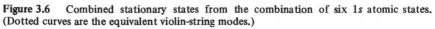

Atom A   Atom B   Atom C   Atom D   Atom E   Atom F

Figure 3.6 Combined stationary states from the combination of six $1s$ atomic states. (Dotted curves are the equivalent violin-string modes.)

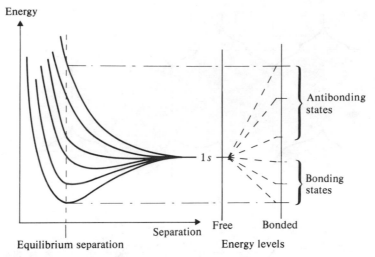

Figure 3.7 Energy versus separation and energy levels for combined 1s states of six hydrogen atoms.

highest-energy ground state into the vacant excited state just above. This fact has important consequences when we come to explain electrical conduction.

(c) N hydrogen atoms (where N is very large)

We will now leap from our six-atom 'solid' to one in which we have a situation more like that in a real solid with N atoms, where N is a very large number (of the order of 10^{29} atoms per cubic metre). We cannot draw the stationary states, because there are so many of them. For the 1s shell of hydrogen, there will be N 1s electrons in the ground state and N vacant 1s states; therefore, there will be $2N$ combined stationary states, each capable of occupation by one electron (or N types of state each capable of containing two electrons). By analogy with the six-atom solid, the ground state of the solid will be that in which the electrons will occupy the lower-energy half of the states. We can draw the envelope of the variation of energy with separation and this is shown in Fig. 3.8. The states are so close together that they form a quasi-continuous band of states between the lowest energy state and the highest energy state — or up to the continuum energy when the highest energy state reaches this energy.

This energy-level scheme at equilibrium separation is also shown in Fig. 3.8 for solid hydrogen containing N atoms. We cannot draw the individual levels, so the band of energy levels corresponding to the allowed states is called an 'energy band'. In this example, the lower half of the band is occupied, the upper half is vacant, and the states are so close that for all practical purposes all values of energy are allowed within the band.

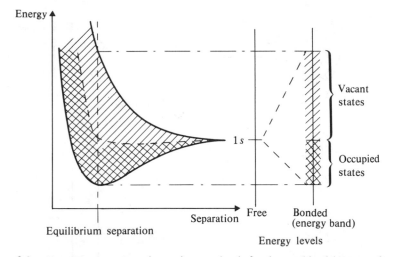

Figure 3.8 Energy versus separation and energy levels for the combined 1s states of a very large number of hydrogen atoms.

(d) N sodium atoms

Hydrogen was chosen for our first example of a solid because it has the simplest energy-level scheme, and, although solid hydrogen is not of much use, the concepts we have introduced also apply to more complex solids. For many-electron atoms, we would expect each energy level to split into a band of combined states when a solid is formed. But combined states only form when the electrons interact with each other, in other words, when the electrons form bonds. The electrons that do this are the valence electrons. The inner electrons are tightly bound to their atoms and do not interact with each other. They therefore sit on identical, degenerate, atomic energy levels, with no way of interacting with electrons from other atoms.

Take, for example, sodium. This has an outer electron configuration of $2p^6 3s^1$, so the valence electron configuration is very like that of hydrogen. The variation of energy with separation, and the band structure of sodium, is shown in Fig. 3.9. This is no longer schematic, as you can see that the energy scale is calibrated in electron volts and the depth scale in angstroms.

The $2p$ level does not split at the equilibrium separation; this tells us that the $1s$, $1p$, $2s$, and $2p$ electrons are tightly bound to the nucleus and take no part in the bonding. The level containing the $3s$ valence electron splits in just the way described for the $1s$ electron of hydrogen.

A band of $2N$ states is formed from the one occupied state and one vacant state of the $3s$ shell. We would say by analogy with the previous example that the ground state of the solid is where the lower half of the $3s$ band is occupied

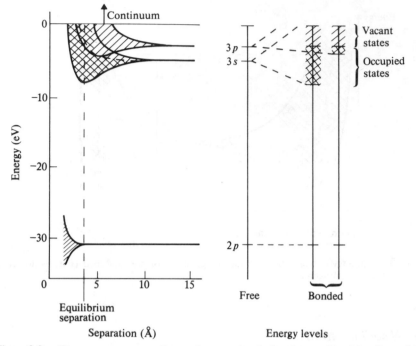

Figure 3.9 Energy versus separation and energy bands for the combined $2p$, $3s$, and $3p$ states of solid sodium.

and the upper half is vacant. Before we can assume this, we have also to consider what happens to the $3p$ states, which are vacant in the atom (with the electron in the ground state) but are very close in energy to the $3s$ states. These also split, to form a band which overlaps the $3s$ band to such a degree that some of the lowest-energy combined states in the $3p$ band come into the lower half of the $3s$ band. As a result, they can be occupied by electrons with the solid in the ground state. For this reason, the $3s$ band is not quite filled to halfway, as some electrons occupy the bottom levels of the $3p$ band. When this overlap occurs, the electrons do not distinguish between states in the $3s$ band and states in the $3p$ band. The overlap produces a super-band with even more states between the lowest-energy state in the $3s$ band up to the continuum.

The band structure at equilibrium separation is shown in Fig. 3.9. For the purposes of clarity, when we show the band structure (the energy level scheme of the solid) we separate overlapping bands, but you must always bear in mind that this separation is just for purposes of illustration.

3.5 The band structure of the elements

If we now look back at Table 1.7, our cross-section of the periodic table, not only do we have examples of all important types of bonding and crystal structure but also we have examples of the important types of band structure. As we have already said, the band structure departs from the energy level scheme of a single atom mainly where the valence-electron states (both occupied and unoccupied) are concerned. For van der Waals bonding and ionic bonding, we have effectively closed shells and so we would not expect energy bands to form; the energy levels would remain as they were in the isolated atom.

Valence electrons interact and, therefore, a band structure does develop for metallic and covalent bonding; it is fundamentally different for the two cases.

(a) Metallic bonding

We have already studied the band structure of sodium. This is representative of all the alkali metals, where one has just one s state occupied.

An interesting case is magnesium, which is next to sodium in the periodic table, with an outer electron configuration of $2p^6 3s^2$. The band structure of magnesium is shown schematically in Fig. 3.10(a). As we have a full $3s$ orbital in an atom of magnesium, we would have a completely full $3s$ band if it were not for the overlap of the vacant $3p$ band, which provides states capable of occupation by electrons with the solid in the ground state. Magnesium is representative of all the Group 2a metals.

The Group 3a metals (aluminium is our example) have the $s^2 p^1$ configuration, so that one would expect the s band to be full but the p band to be only partially filled.

The third class of metals in the periodic table are the transition metals (and also the lanthenides and actinides). We will take as our examples to represent the band structure of transition metals two very important elements: iron and copper. The band structure of iron is shown in Fig. 3.10(b). The valence electron configuration of an atom of iron is $3d^6 4s^2$. However, when we inspect the band structure of iron in solid form, we see that there is complete overlap of the $3d$ band by the $4s$ band, and the $8N$ electrons (eight valence electrons per atom) are shared between the $12N$ states (two s states and ten d states per atom) in such a way that neither the $3d$ band nor the $4s$ band is completely full.

It is the almost full $3d$ band that is the key to the unique ferromagnetic properties of iron and the two other Group 8 metals, cobalt ($3d^7 4s^2$) and nickel ($3d^8 4s^2$).

Copper, the next element along from nickel, has the valence-electron configuration $3d^{10} 4s^1$, with a band structure (Fig. 3.10(c)) that differs very little from that of the Group 8 metals. The only important difference is that the $3d$ band is completely full. The $4s$ band is still partially filled, as we have $11N$

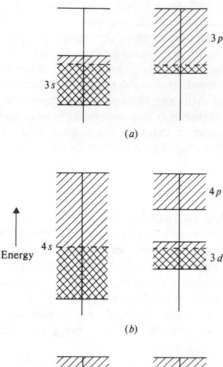

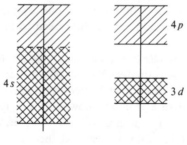

Figure 3.10 Valence-electron energy bands for (*a*) magnesium, (*b*) iron, and (*c*) copper.

electrons to share betweeen $12N$ states. This structure is representative of all the Group 1b metals and similar to the alkali metals of Group 1a.

The Group 2b metals have full *s* and *d* states in atomic form, but, just like the Group 2a metals, the overlap of the *p* and *s* bands means that we have partly-filled upper bands.

(b) Covalent bonding

For covalent bonding, the valence electrons of neighbouring atoms form a cooperative state, which is called a molecular orbital. The atoms in a covalently-bonded crystal are linked by a network of molecular orbitals. Therefore, all the valence electrons can interact with one another and so, as in the case of metals, a band structure must form.

The type of band structure resulting from this form of interaction can best be understood with reference to the energy-level diagram given when we were discussing the covalent bonding of silicon (Fig. 1.11). The energy-level scheme, which is really only relevant to a single atom, tells us that, as the interatomic separation decreases, the two filled $3s$ states, two filled $3p$ states, and four vacant $3p$ states combine to form eight hybridized states, four filled and four vacant. As the interatomic separation decreases further, these hybridized states split into four filled bonding states and four vacant antibonding states. As the atomic separation decreases in a solid of N silicon atoms, it is more realistic to say that $2N$ atomic $3s$ states and $6N$ atomic $3p$ states ($2N$ filled and $4N$ vacant) combine to form a *band* of $8N$ hybridized states; this then splits to form a *band* of $4N$ bonding states and a *band* of $4N$ antibonding states. Bands are formed as all the valence electrons interact with each other through a chain of covalent bonds, and so the exclusion principle holds. This behaviour is common to all the Group 4a elements and a schematic representation of the variation in band structure with interatomic separation is shown in Fig. 3.11. The band diagram for diamond, silicon, and germanium at the equilibrium separation in the solid is also shown.

One can see that, in the ground state, the band of bonding states is full whilst the band of antibonding states is empty. This is generally true for covalent solids in their normal valency configuration.

The band of bonding states is normally called the valence band as it contains the valence electrons. The band of antibonding states is called the conduction band, because an electron has to be excited into an antibonding state (removed from the covalent bond) before it can take part in electrical conduction.

The point of equilibrium separation, on the schematic diagram in Fig. 3.11, changes as we progress down the periodic table. It is the magnitude of the gap in allowed energies (called the band gap) between the valence band and the conduction band that is important in determining whether a covalently-bonded solid is an insulator or a semiconductor. Diamond, which is a good insulator, has a very large band gap in atomic terms (5 eV) whilst the semiconductors silicon and germanium have small band gaps (1 eV and 0.7 eV, respectively). The band gap for the Group 4a metal tin is either close to zero or even negative (i.e., overlap in the hybridized region).

3.6 Band structure and electrical conduction

We can now use the band structure to explain the electrical nature of solids.

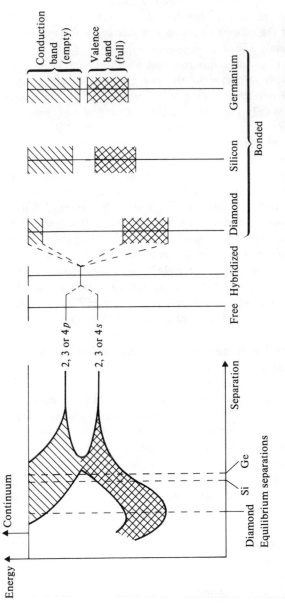

Figure 3.11 Energy versus separation and energy bands for the valence electrons of diamond, silicon, and germanium, the Group 4a elements.

For metals we have the situation that the valence band is never completely full. There are always vacant states for one of the following reasons:

1. the valence band derives from an unfilled set of atomic states (e.g., sodium and aluminium),
2. there is overlap of the valence band with a band of states derived from a normally empty set of atomic states (e.g., magnesium),
3. we have a hybridized band with vacant antibonding states forming a continuous band with the occupied bonding states (e.g., tin).

An electric current flows through a solid only if there are electrons which can drift under the influence of an electric field. This occurs when the energy of the electrons can be increased slightly in the direction of the electric field. As we know, the electron can only have certain allowed energies defined by the set of stationary states available. Thus, the electron energy can only be raised if there are vacant states available at the required energy. As illustrated in Fig. 3.12(a), this occurs for the highest-energy electrons in a partially filled band. Within the band, there is a quasi-continuous set of vacant states at only slightly higher energies than the highest-energy electrons. These can be occupied when an electric field is applied to the solid.

One can see that the energy that must be given to the electron is very small compared with the total energy, so an electric current can be thought of as a very gentle drift superimposed on violent random motion.

Metals are conductors because they have vacant states in the valance band.

Covalently-bonded solids (if all the valence electrons take part in bonding) have a completely full valence band (the bonding states) and so, in the ground state, there are no vacant levels that could be occupied by electrons excited by an electric field.

The electrons of lower energies within the valence band cannot be raised in energy because there are no vacant states within the band. A swop in position of two electrons at different energies within the band would be possible, but, as electrons are indistinguishable, we would have no change in electron energy distribution and so no electrical conduction. The band gap is normally too large for electrons to be excited to occupy states in the vacant conduction band (the antibonding states) by application of an electric field.

Thus, covalently-bonded solids are usually insulators because they have no vacant states in the valence band.

We have talked of a solid being in the 'ground state', and this implies that there is no thermal energy, i.e., the solid is at the absolute zero of temperature. At high temperatures, we find that some covalently-bonded solids conduct electricity. These are the third — and electrically very interesting — class of solids: the semiconductors.

Thermal energy arises from the vibration of atoms about the equilibrium separation. This variation in atomic position distorts the combined electron states, resulting in a change in energy of the electrons. At any instant of time,

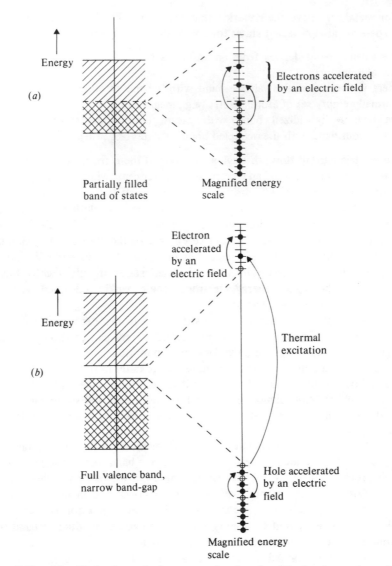

Figure 3.12 Magnified view of the energy bands of solids in the region of the highest-energy electrons. (*a*) For metals (partially filled valence band); (*b*) for semiconductors (filled valence band, narrow band-gap).

some electrons will have gained energy from the thermal vibrations. Another way to look at this is to say that the covalent bonds are bent by thermal vibrations and, occasionally, a bond will break and an electron will be released.

In terms of the band-diagram model, we can say that a small proportion of electrons can occupy excited states as much as 2 eV above the state that would be occupied by the highest-energy electrons at absolute zero (the top of the valence band). Normally, for a covalently bonded solid, this energy would fall in the band gap; however, if the band gap is small, say 1 eV or less, and the temperature is high enough, there will always be a few electrons free of the bonds. These will occupy states at the bottom of the conduction band (Fig. 3.12(b)). These electrons can, therefore, be raised in energy by an electric field and the solid can conduct electricity. There are also levels at the top of the valence band vacated by electrons excited to the conduction band. These vacant levels enable electrons in the valence band also to take part in electrical conduction, in a way analogous to metallic conduction. Because there are very few vacant levels in a vast number of electron occupied levels, conduction is treated as though it were by movement of the vacant states in the opposite direction to the field. The vacant state is an electron missing from a bond and, thus, charge neutrality in this region does not hold; hence, the bond that is missing an electron is positively charged, and, as electrons hop into the vacancy excited by the electric field, the vacancy moves in the opposite direction. The positively charged electron vacancy is called a 'hole'. The two different types of electrical conduction (both really by electrons in different ways) distinguish semiconductors from metals.

There are never as many electrons (or holes) available for conduction in semiconductors as in metals. This is because, of all the possible thermally-excited electrons, only the few of very highest energy can reach available states in the conduction band. Thus, semiconductors have a much lower conductivity than metals; hence the name.

To summarize, we can say that semiconductors are covalently-bonded solids that will conduct electricity at high temperatures because they have a small energy gap between the valence band and the conduction band.

It is worth noting here that ionically-bonded solids also conduct electricity at temperatures near to their melting point. This is not electron conduction but is due to movement of the ions themselves as mobile defects in the crystal structure.

3.7 Free electrons

The model that we have just developed, which led to the band diagram, has been arrived at by studying the way that stationary atomic states for bound electrons are modified in a solid. This is the tight-binding model, and we have seen how useful it is in explaining the main electrical properties of solids. However, this model has its limitations and in order to explain many detailed effects – such as electrical resistance of metals and the mobility of electrons, emission of light, and electrical oscillations in semiconductors – we need to know how the momentum of an electron and the density of electron states varies with energy.

We will briefly describe a model that enables us to understand these phenomena. In this, we consider the valence electrons to be completely free within the solid and then consider the effect of the atomic closed-shell cores.

This is the loose-binding model, and, as you will see, the approach is just the opposite to that for the tight-binding model.

3.8   Density of states

We will determine the properties of a metallic solid using a model in which the valence electrons are free, forming an electron gas confined only by the surface of the solid. This approach is not a new one. It was first proposed by the German physicist Paul Drude in 1900 and was used successfully by the Dutch physicist Hendrik Lorentz to explain electrical and thermal conduction and the constant ratio between the two conductivities.

Their model was remarkably successful – considering that it was based on the false assumption that the electron gas behaved like a classical gas. We now know that classical mechanics, such as Newton's laws, are just an approximation to more accurate mechanics, based on quantum theory, and that the classical approximation can only be applied when the system to be described is large. This is not the case for electrons, to which the ideas of quantum mechanics must be applied, in particular the concepts of allowed stationary states, energy levels, and the Pauli exclusion principle. The classical model failed to explain a number of phenomena, such as the very small electron specific heat (the increase in electron energy on raising the temperature of the solid), for this reason.

In this treatment of the free-electron model, we will use the concepts of quantum mechanics and, as a starting point, we will once more use the idea of violin-string stationary states, as introduced in Chap. 1. The electron gas, confined to fall to zero amplitude at the surface of the solid, will form a quantized set of stationary states, and, therefore, there will be a set of energy levels analogous to the set derived in Chap. 1, where we found that the energy $E_n = h^2 n^2 / 8ml^2$. In the case of an electron gas, l is equivalent to the linear dimensions of the solid. (Remember that this simple derivation is for a one-dimensional oscillation.) The energy spacing between states is inversely proportional to l^2 and, as l is very large on an atomic scale, this means that the energy spacing of the states is very small; hence, we have almost a continuum of states over all values of energy. This is why the classical theory worked so well.

It is now important to determine how many states there are per unit increase in energy.

In one dimension, it appears as though the states get further and further apart as the energy increases because $E_n \propto n^2$; this would imply that the density of

Energy, E

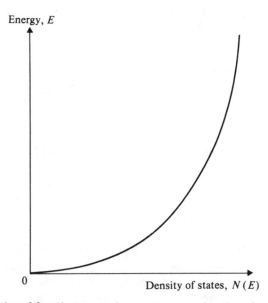

0   Density of states, $N(E)$

Figure 3.13   Density of free-electron stationary states as a function of energy. (Energy is plotted vertically to be consistent with the energy-level diagrams.)

states (the number of states $N(E)$ in the energy interval dE), gets smaller as E increases. We must again remember that we are working in three dimensions, and so we must think of quantum numbers of waves directed not only in the x but also the y and z directions. These waves will have quantum numbers n_x, n_y, and n_z. The larger the quantum numbers become, the more quantum states are available.† In fact, it can be calculated (although we will not do so here) that the number of states at $n_x = n_y = n_z = n$ is proportional to the surface area of a sphere of radius n. It works out that the density of states $N(E)$ is proportional to the volume of the solid (V) times the square root of energy:

$$N(E) \propto V \times \sqrt{E}.$$

A schematic plot of density of states versus energy for a free-electron gas is shown in Fig. 3.13. Energy is plotted vertically to be consistent with the energy-level and band diagrams. The density of states obeys a parabolic relationship.

It is interesting to note that the increase in the density of states with the increase in the volume of the solid can be predicted from the Heisenberg

---

†Remember that, when we presented the energy level scheme for a hydrogen atom at the beginning of this chapter, we observed that the energy separation between states decreases as energy increases and vibrations become more complex.

uncertainty principle. An increase in volume implies an increase in uncertainty of the position of an electron, so we can know the momentum of the electron more accurately and to do this there needs to be a higher density of states.

3.9 The Fermi energy

We have established that the free electrons form a quasi-continuous band of energies from zero energy upwards, with the density of states increasing as the energy increases in the way just described.

How will these levels be occupied by electrons? Just as in the case of energy levels for an atom, we must apply the Pauli exclusion principle, which predicts that in any system of interacting electrons (which in this case is the whole solid) only one electron can occupy each of the available states.

In the ground state, the first of the free electrons (about 10^{29} per m^3) will occupy the lowest energy state; subsequent electrons will have to fill higher and higher energy states until the last free electron in the solid, let us say number 10^{29}, will have to occupy a very high energy state. This is a remarkable consequence of the exclusion principle, because we have assumed that the solid is in the ground state, implying a temperature of absolute zero.

The energy of this highest-energy occupied state in the solid at absolute zero of temperature is called the Fermi energy E_F (the energy level is called the Fermi level).

By a simple integration of the expression for the density of states from 0 to E_F (not done here), it can be proved that the Fermi energy is proportional to the number of free electrons per unit volume, N/V, to the power of two thirds:

$$E_F \propto (N/V)^{2/3}.$$

Thus, E_F, unlike the density of states, is independent of the volume of the solid. This can be understood if you compare a solid with one of half the size; there would be half the number of electrons in the smaller one, but the density of states would also be halved and E_F would be unaltered.

When the solid is at high temperatures, the distribution of electrons is altered. We have already implied this in Sec. 3.6, when considering electrons that reach the conduction band of semiconductors. Some electrons near to the Fermi energy can occupy states at a higher energy than E_F, and so some of the levels immediately below E_F will be vacant. Electrons at energies well below E_F cannot be excited because there are no vacant levels for them to occupy. The distribution of electrons in the region of E_F is governed by the universal expression that is used when determining the variation in the numbers of excited electrons as a function of temperature: $\exp(\Delta E/kT)$, where ΔE is an energy (which in this case is $E - E_F$), k is Boltzman's constant, and T is temperature. The full expression for the probability of an electron having energy E as a function of energy $P(E)$ is given by

$$P(E) = [\exp\{(E - E_F)/kT\} + 1]^{-1}.$$

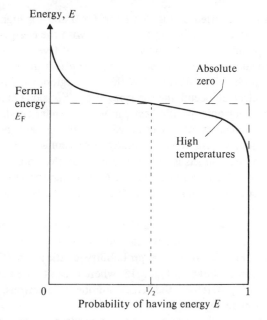

Figure 3.14 The probability of an electron having energy E as a function of E.

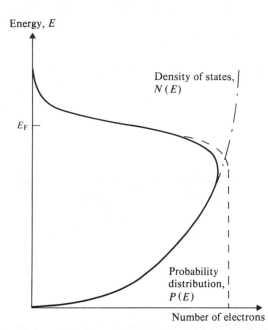

Figure 3.15 The number of electrons as a function of energy, obtained by multiplication of the density of states by the probability distribution.

This expression is plotted in Fig. 3.14 for $T = 0$ and for high temperatures. Energy is again plotted vertically to be consistent with the energy-level and band diagrams. The diagram for $T = 0$ tells us only that all levels are occupied up to E_F and none at higher energies. We can see that the effect of heating is simply to round off the probability distribution in the region of E_F, so that the probability of an electron occupying the state at $E = E_F$ is one half.

The very small electronic specific heat can now be understood, because, when the temperature of a classical gas is raised, the *average* energy of the whole distribution is raised, with a consequentially large change in energy. When the temperature of an electron gas is raised, the bulk of electrons occupying lower energy states is not affected and the average energy changes very little. Only the few electrons occupying states near to the Fermi energy can be excited to occupy higher energy states.

The actual number of electrons as a function of energy for a free-electron gas obeying the quantum-mechanical laws can now be obtained by multiplying the density of states (Fig. 3.13) by the probability distribution (Fig. 3.14). The resultant distribution is shown in Fig. 3.15, where it can be seen that the highest density of states, and, therefore, the largest number of electrons, is at an energy just below the Fermi energy.

This model works well for metals, but, as we know from Sec. 3.5(b), for covalent bonding we have a band gap of forbidden energies above a full valence band, so the electrons are not able to occupy levels just above E_F at high temperatures. In fact, E_F lies at the exact centre of the band gap, so electrons cannot occupy levels just below E_F either. It is not really surprising that the free-electron model falls down in solids where the electrons are not free but are trapped in covalent bonds. However, it turns out that we can modify the free-electron model to predict the band gap by considering the scattering of the free electrons by the atomic cores (or ions) arranged in their regular crystalline array.

3.10 Interaction of free electrons with the crystal structure

It is very useful in understanding many phenomena in solids, to study how the momentum of the electron varies with energy. It is in terms of this relationship that we will describe the interaction between free electrons and the crystal structure.

In Chap. 1, we derived this relationship for the violin-string states, namely

$E = p^2/2m.$

It is more convenient to consider momentum in terms of wavelength, and the de Broglie relationship tells us that $p = h/\lambda$, so we can consider momentum as

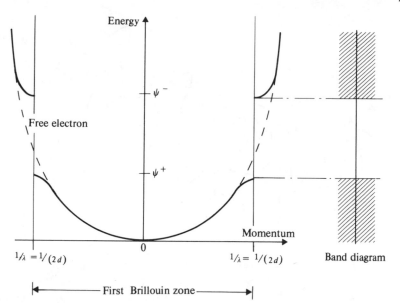

Figure 3.16 Energy versus momentum for free electrons and for electrons in a one-dimensional crystal of atomic spacing d showing the energy gap due to the resonant scattering condition.

the reciprocal of the wavelength.† For a free electron, energy is proportional to the square of the momentum and, therefore, is proportional to $(1/\lambda)^2$.

This parabolic relationship is shown as the dotted line in Fig. 3.16. We only consider the momentum in one dimension, but, as this can be in either direction, the graph shows negative values.

The free-electron travelling waves interact with the atomic cores by being scattered in all directions. In general, this scattering is random; the electron waves scattered from one atom interfere destructively with those scattered by neighbouring atoms and all the scattered waves cancel each other out, leaving only the wave propagated forward that is in phase with the incident wave. This is just the wave that would exist if no atoms were there in the first place; the electrons act as though they were truly free.

We have already mentioned, in Chap. 1, that electrons can be diffracted by a crystal lattice, just as light is diffracted by an optical grating. The general case for diffraction is shown in Fig. 3.17(a). This is not only true for an electron beam in a microscope, but is also true for valence electrons in a solid. (Fig. 3.17(a) is a repeat of Fig. 1.2.)

---

†In more mathematical treatments, the momentum of a wave is described using the wave vector **k**; this has the magnitude of $1/\lambda$ or $2\pi/\lambda$, depending on which definition is more convenient, and also describes the direction of propagation of the wave.

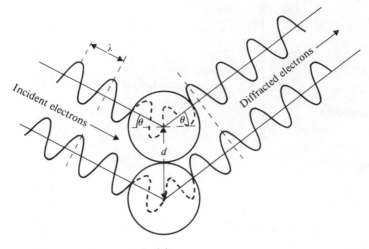

(a)

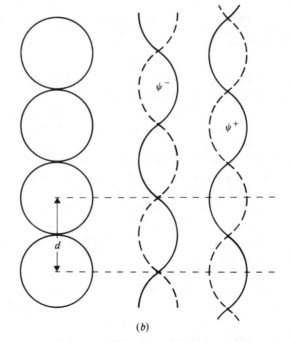

(b)

Figure 3.17 The necessary condition for diffraction of electrons and X-rays in crystals. (a) The general case: $n\lambda = 2d\sin\theta$. (b) The one-dimensional case: the resonant condition and creation of standing waves. $\theta = 90°$, $n\lambda = 2d$.

Diffraction occurs when the interference condition is satisfied. In one dimension, this is given by $n\lambda = 2d$, where d is the spacing between the scattering crystal planes and n is an integer.

When $n\lambda = 2d$, the scattered electron waves interfere constructively (as illustrated in Fig. 3.17(b)), energy cannot be propagated, and we have a resonant condition. The electron wave is no longer a travelling wave, but becomes two standing waves, as illustrated.

It is not obvious from our model why *two* standing waves are formed. One could say that one standing wave (ψ^+) is the resonance of electrons located in covalent bonds and, therefore, within the atoms, whilst the other standing wave (ψ^-) is the resonance of free electrons which, by definition, cannot enter the atoms and occupy the spaces between them.

One standing wave (ψ^+) has a maximum probability density at the atomic cores and, therefore, has a low potential energy. The other standing wave (ψ^-) has a maximum between the atoms, resulting in a higher potential energy. An electron cannot have an energy between these two values. This concept is the same as that of the band gap, but now developed from the free-electron model.

The resonant condition at $\lambda = 2d$ causes a discontinuity in the energy versus momentum diagram. The relationship is plotted as a solid line in Fig. 3.16. The relationship starts to depart from the free-electron parabola as we approach the resonant condition, and there is a jump in energy at resonance. The corresponding band diagram is also shown in Fig. 3.16.

It seems at first rather strange that the tight-binding model, based on bound electron states, gives the same result as the loose-binding model, based on the scattering of free electrons. It is not so strange if we realize that, in the resonant condition for free electrons, the electron wavelength is an integral subdivision of the atomic diameter, and is a standing wave. This is just the condition for a combination of atomic stationary states.

The band diagram shown in Fig. 3.16 can be understood in relation to that developed earlier if we take the zero of energy (zeros are always conveniently chosen) as the centre of the valence band; then, the first resonance, at $\lambda = 2d$, corresponds to the energy gap between the valence and the conduction bands, the second resonance, at $\lambda = d$, corresponds to the next band gap — at the top of the conduction band.

One must remember we have just considered a one-dimensional model, but, as the spacing between crystal planes varies with direction, so the momentum for resonance also varies with direction in the crystal structure. The first zone, between a momentum of $(1/\lambda) = (1/2d)$ and $|-(1/\lambda)| = (1/2d)$ (for a wave travelling in the opposite direction), has a beautifully regular shape in three dimensions, reflecting the symmetry of the reciprocal of the crystal-plane spacing. The surface of this so called Brillouin zone represents the value of the momentum at the band edge.

Another way of representing the variation of energy with momentum is to move the zero of momentum to the momentum for resonance. We can consider

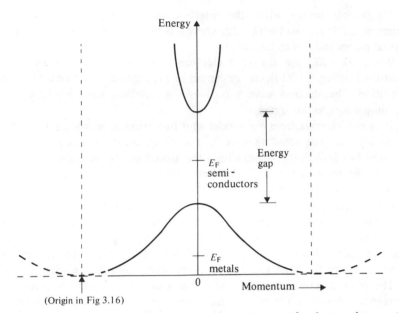

(Origin in Fig 3.16)

Figure 3.18 The periodic zone scheme; energy versus momentum for electrons in a crystal with the zero moved to the resonant condition. Because of periodicity of properties the relationship is symmetrical about this point.

the diagram to be symmetrical about this point (electrons can have momentum in the opposite directions in both the conduction and the valence bands) and so arrive at the representation shown in Fig. 3.18. This is called the periodic zone scheme.

3.11 Combination of models

(a) Fermi energy and the energy versus momentum diagram

We have now introduced all the basic concepts that are necessary to understand most of the physical properties of solids. We need now to show how some of the features of the models can be combined.

In Fig. 3.18 is shown the position of the Fermi level on the energy—momentum relationship for metals and for covalently-bonded solids. For metals, the Fermi level is usually well away from the band edge, where the energy—momentum relationship is almost identical to the free-electron parabola. So, electrons in metals will behave like free electrons. In three dimensions, the value of momentum at the Fermi energy forms a surface (called the Fermi surface). For free electrons, momentum at the Fermi energy is the same in all directions, so the Fermi surface is a sphere.

For covalently-bonded solids, the Fermi energy lies at the centre of the band gap and the Fermi surface is related to the crystal structure.

For semiconductors, conduction is by electrons excited to the bottom of the conduction band or by holes left at the top of the valence band.

At these values of momentum, we are very close to a resonant condition and well away from the free-electron energy—momentum relationship. Although electrical current can be carried by electrons and holes, they do not behave like free electrons. In order to use equations really applicable only to free electrons, we say that the charge carriers have an effective mass which is related to the departure of the energy—momentum relationship from the free-electron parabolic relationship. This is not a true change of mass, but just reflects the different way electrons behave at values of momentum close to the resonant condition.

(b) The density of states and the band diagram

We have seen how the density of states increases with energy for free electrons (Fig. 3.13). The variation of the density of states within a band is related to this and is illustrated in Fig. 3.19. For the lower half of the band, the density of states behaves like the free-electron density, with zero energy at the bottom of the band. For the upper half of the band, it is more useful to think of holes as the charge carriers, with a density of states behaving like the inverse of the

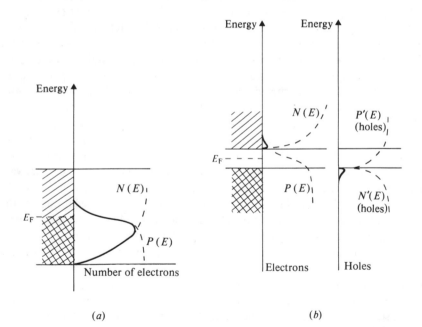

Figure 3.19 Combination of the band diagram with the free-electron model, showing the number of electrons as a function of energy for (a) metals, (b) semiconductors. In the semiconductor case, relations for holes in the valence band are the mirror image of those for electrons in the conduction band: $P'(E) = 1 - P(E)$.

free-electron density with zero energy at the top of the band. The density of states increases as we approach the centre of the band. The number of electrons as a function of energy (at high temperatures) is also shown, in Fig. 3.19, for the valence band of metals and for semiconductors. This is found by multiplying the density of states by the probability distribution, as we did for free electrons in Sec. 3.9.

Metals are good conductors because of the large number of electrons available for conduction; this, in turn, is due to the high density of states at the Fermi energy. Semiconductors rely for conduction on states very close to the band edges, where the density of states is very low, so the number of charge carriers is small and conductivity is low.

3.12 Use of models

(a) Doping of semiconductors

One can see from Fig. 3.19 that, for a pure semiconductor, the Fermi level is at the centre of the band gap and the number of holes equals the number of electrons. This balance can be altered by incorporating suitable impurities in the crystal structure. These are called dopants. Phosphorus, which is a commonly used dopant, has the outer electron configuration $3s^2 3p^3$ and, normally, only the p states form bonding states (three bonds), with the s states remaining unaltered. If phosphorus is forced to sit in a silicon lattice, it will be obliged to form a tetrahedron of four bonds by splitting of the energy levels, as shown in Fig. 3.20(a). The s states contribute one electron to the bonding states, leaving the surplus electron in an antibonding state. In the solid, this surplus electron forms a small band of occupied states just in the vicinity of a phosphorus atom. This impurity band is close to the conduction band; the Fermi energy for the whole solid must now lie in the very narrow gap between the impurity band and the conduction band, and so is very close to the edge of the conduction band. As a result, we have a very large increase in the number of electrons in the conduction band. This occurs even at very moderate temperatures because very little energy is needed to excite electrons from the impurity band to the conduction band. The shift in the Fermi energy causes a reduction in the number of holes at the top of the valence band. The value of the conductivity will be greatly increased by addition of a small quantity of dopant. In addition, the type of conductivity will be shifted, so that, in this example, electrons in the conduction band predominate.† A semiconductor doped in this way is called an n-type semiconductor, as the majority charge carriers are negative.

Semiconductors where electrical conduction is mainly by positive holes, p-type semiconductors, can also be made. This is done by using a dopant, such as

---

†The positive charges left in the small antibonding electron band cannot move as they are trapped on the phosphorus impurity atoms, which have no electrons missing from the tetrahedral bonds to neighbouring silicon atoms. (There are no mobile holes.)

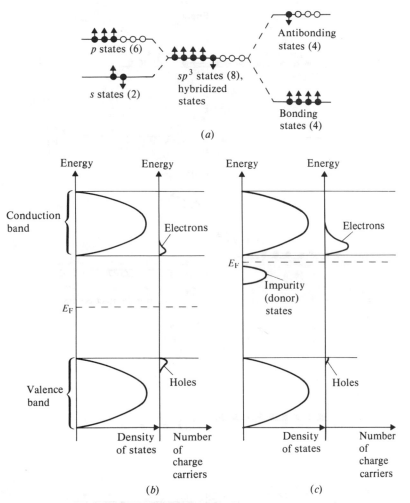

Figure 3.20 The doping of silicon with phosphorus. (*a*) Energy levels; (*b*) intrinsic density of states and number of charge carriers versus energy; (*c*) effect of phosphorus doping on the density of states and number of charge carriers versus energy.

boron ($2s^2 2p^1$), in which we have a vacancy in the bonding states, creating a narrow band of vacant states just above the valence band. This forces the Fermi energy to lie close to the top of the valence band, with a consequent increase in the number of holes at the top of the valence band and decrease in the number of electrons in the conduction band. The electrons excited to the narrow vacant impurity band are trapped on the bonds to the dopant atoms.

86

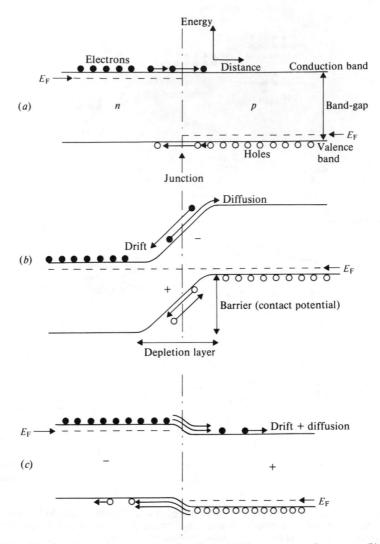

Figure 3.21 The band diagram at a p–n junction. (a) The moment of contact; (b) at equilibrium; (c) forward bias. ●, electrons; ○, holes.

This ability, permitting the engineer to control not only the conductivity but also the nature of the electrical conduction mechanism in semiconductors, has led to the development of many devices. One of the first was an amplifying device using alternating layers; for example, layers doped in the sequence n—p—n, where a small voltage on the centre layer controls a large current passing through the device. This is the transistor.

The band diagram shifts when n- and p-type semiconductors are placed in contact, such that equilibrium is established and the Fermi levels align. This results in a barrier being created between the n- and p-type regions. It is the control of the barrier height which is the secret of transistor action.

The region of the band diagram close to the band gap is shown in Fig. 3.21 for a p—n junction (energy is plotted vertically and distance is plotted horizontally).

Figure 3.21(a) depicts the situation at the moment of contact. Electrons diffuse from a high concentration in the conduction band of the n-type semiconductor into the empty levels in the conduction band of the p-type semiconductor. Similarly, the holes diffuse in the valence band from p-type into n-type. Figure 3.21(b) depicts the situation at equilibrium, the Fermi level is constant across the p—n junction.

The diffusion of electrons and holes across the junction has created a region depleted of charge carriers. The positive donor and negative acceptor ions that must remain in this region create an electric field across the junction (the contact potential) such that a drift current of electrons and holes occurs, equal and opposite in sense to the diffusion currents. There is no net current flow across the junction.

Figure 3.21(c) depicts the situation where a negative potential is applied to the n-type side of the junction (forward bias). The barrier to current flow is now removed, drift and diffusion are in the same direction, and current flows across the junction.† No current will flow if a voltage is applied to make the n-type positive (reverse bias), as the only effect will be to make the depletion layer wider and the potential barrier higher. The p—n junction will only pass current when forward biased. The rectifying property of the p—n junction is used in a variety of ways to create a vast number of different semiconductor electronic devices.

Nowadays, large-scale integrated circuits are made, where each circuit is the size of a pinhead and contains, typically, ten thousand transistors, resistors, diodes, and capacitors, all formed on one piece of single-crystal silicon by processes such as doping, oxidation of selected areas, and laying down thin metal films to form the necessary interconnections. A circuit of this type is shown in Fig. 2.13.

(b) Work function and the emission of soft X-rays from metals

We can used the relationship between the number of electrons and energy, shown in Fig. 3.19(a), to explain the energy of electrons and X-rays emitted from metals.

---

†Think of the electrons as ball bearings rolling down the slope on the bottom of the conduction band, and holes as bubbles floating up to the ceiling formed by the top of the valence band.

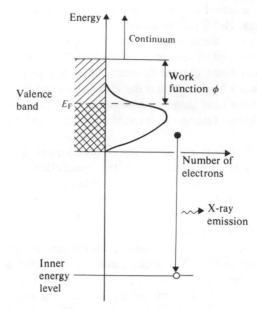

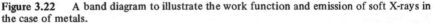

Figure 3.22   A band diagram to illustrate the work function and emission of soft X-rays in the case of metals.

The work function of a metal is the energy required for electrons to escape from the surface of the metal. We can see, with the aid of Fig. 3.22, that this will be the energy difference between the Fermi energy and the continuum energy.

When electrons are promoted to excited states, they stay there some time but, eventually, decay to the ground state. The energy saved thereby is given out as light of a wavelength equivalent to the difference in energy between the excited state and the ground state. We will describe this phenomenon in more detail in Chap. 5. It is interesting to note that, for metals, where an electron vacancy has been created in an inner shell by electron bombardment, we can have a continuous distribution of soft X-rays emitted over a range of wavelengths instead of emission at one well-defined wavelength.

This can be explained easily with our model, because, in these cases, the electrons decay to fill the vacancy from the states in the valence band, and so the soft X-ray spectrum follows the electron distribution in the band. In fact, this phenomenon can be considered as direct evidence of the validity of the free electron model which predicts the density of states and the Fermi energy.

(c) The emission of light by semiconductors

The concept of emission of radiation caused by the decay of electrons also

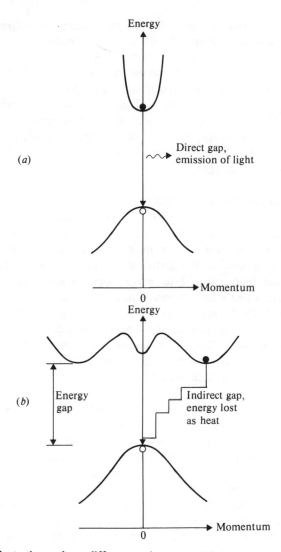

Figure 3.23 Illustrating the difference between (*a*) direct-gap semiconductors, (*b*) indirect-gap semiconductors.

applies to semiconductors; in this case, light in the visible range is emitted by transitions across the band gap.

If an excess of electrons are excited into the conduction band by being injected across a p–n junction, they will decay (recombine with holes), giving out light equivalent in wavelength to the band gap. This is shown in Fig. 3.23(*a*), using the periodic zone version of the energy—momentum diagram. This works well for the compound semiconductor gallium arsenide and is the basis of the light-emitting diodes used as displays in pocket calculators, but it does not work for silicon. The reason why we cannot induce silicon to emit light is that, because of fine detail in the crystal structure, the minimum energy in the conduction band is not at the same momentum as the maximum energy of the valence band (Fig. 3.23(b)). As a result, in order to decay, not only does the electron have to lose energy, but it also has to change momentum. This is only possible by several collisions with atoms, losing energy and momentum in the form of heat rather than light.

(d) Electrical resistance in metals

Electrical resistance in metals (Ohm's law) can be explained by using the one-dimensional energy—momentum diagram first shown in Fig. 3.16. This is reproduced in Fig. 3.24(a) with an indication of the Fermi energy. When an electric field is applied to the metal, the electrons with momentum in the direction of the field gain a little energy and the electrons travelling in the opposite direction lose a little energy. This is shown in Fig. 3.24(b). This change in momentum takes the highest-energy electrons (aligned with the field) closer to the band edge. The momentum at the band edge is determined by the spacing of the crystal planes d. But d does not remain constant; because of thermal vibrations, there is a range of possible values of d. The range is small at low temperatures but large at high temperatures. Crystal defects also cause localized variations in d.

Electrical resistance arises from the fact that a proportion of the electrons accelerated by the electric field meet a region of the crystal where the spacing is such that the electron has an energy in the band gap, creating a resonant condition. The electron is strongly scattered and acquires momentum in the opposite direction, so reducing the electric current arising from the drift of electrons with the applied field.

Ohm's law arises because the greater the electric field, the greater the shift of the energy—momentum distribution in the direction of the field; therefore, the larger the number of electrons that occupy regions of the crystal where they are strongly scattered.

The linear increase of resistance with temperature (Fig. 3.24(c)) for metals at high temperatures results from the increase in the spread of values of d (with consequently more scattering of the conduction electrons) due to the greater amplitude of the thermal vibrations. The residual resistivity at very low temperatures is due to crystal defects, a 'frozen-in' variation in d.

3.13 Superconductivity

For a normal conductor, there is always a residual resistance at low temperatures. However, for some metals, at temperatures below 10 K, there is a

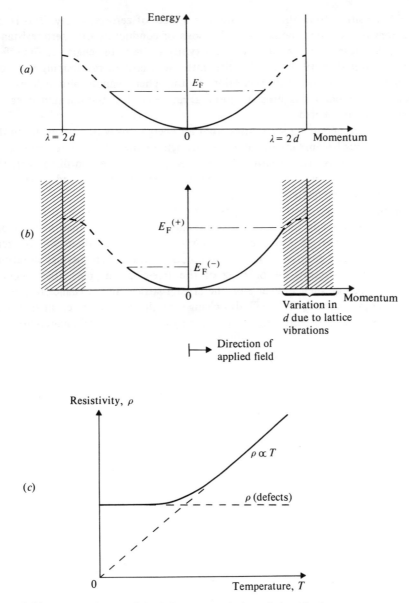

Figure 3.24   Ohm's law explained in terms of the relationship between energy and momentum of conduction electrons. (*a*) No electric field; (*b*) electric field applied: (*c*) the variation of resistivity with temperature.

rapid transition from the normal state to a state of zero resistivity. This is the phenomenon of superconductivity. The idea of conductors with zero resistance (and, therefore, zero loss of power) is very attractive to the engineer. The effect was discovered in 1911, but the difficulties and cost of maintaining very low temperatures has delayed application of this phenomenon, although experimental superconducting magnets, motors, and electrical transmission lines are now being constructed.

The superconducting state arises from an energy saving that occurs when the electron spins become ordered in pairs throughout the solid by interaction with the crystal lattice. One electron interacts with the lattice and distorts it, the second electron can use this distortion to lower its energy, and so the two electrons interact via the crystal lattice.

If the electrons drift in an applied field, the interaction with the lattice distorts the lattice in such a way that the band edge moves so that the electrons are not scattered. If the electrons cannot be scattered, there will be zero resistance. The transition to normal conduction arises when thermal vibrations of the lattice destroy the ordering of the free-electron spins; they become randomized, lose their influence on the crystal lattice, and resistance reappears. An external magnetic field will also change the direction of electron spin, and therefore at a sufficiently high value of magnetic field superconductivity will dissappear.

4. Magnetism

4.1 Introduction

In Chap. 3 we introduced the phenomenon of electrical conductivity. A second important consequence of the way the valence electrons behave in a solid is the phenomenon of magnetism.

We will treat this subject in the same progressive way that we approached the subject of electrical conductivity. We will start by introducing the concept of magnetic forces and fields, go on to describe the magnetic properties of single atoms, and then look at the effect of having large assemblies of atoms — first for the general case (paramagnetism and diamagnetism) and then for the special case of ferromagnetism. The chapter ends with a description of the engineering aspects of ferromagnets, including three case histories showing how ferromagnetic properties are tailored to suit particular applications.

4.2 Electrical forces

Electrical forces are very large on our scale of things. The only thing that prevents matter tearing apart is that these forces are finely balanced; charge neutrality, for example, is pretty vigorously observed. One can appreciate this fact if we consider two people standing half a metre apart, each with one per cent more electrons than protons. There will be an electric repulsive force between them given by Coulomb's law,† and the magnitude of this repulsive force would be great enough to lift a mass equivalent to that of the entire earth!

These electrical forces are the driving forces behind atomic bonding. The quantum effects, the large energy needed to confine an electron to a small volume, prevent the complete collapse of matter. It is the balance between electrical forces and quantum effects that determines the detailed nature of matter.

---

† This is the 'inverse square' law. The repulsive force F between like charges q of separation r is given by $F \propto q^2/r^2$.

The electrical force has two components. One of these arises from the interaction of static electrical charges. This force is described by saying that a point charge creates an electric field which acts on other point charges. The second component depends on the *motion* of the electric charge. This is the magnetic force. A magnetic field is created when an electric charge is in motion.

The magnetic field, in fact, arises from a tiny correction that needs to be made to the electrical forces due to relativistic effects. Strange things happen when we approach the speed of light. This is another case where classical Newtonian mechanics falls down, but in this case it is because things (velocities and distances) get very big.† The correction really is tiny, of the order of v^2/c^2, where v is about 1 m s^{-1} for a current in a wire (remember we only have a slow drift with the applied field) and c, the velocity of light, is 10^8 m s^{-1}. We need to correct to the order of one part in ten thousand million million. It is only because electrical forces are so vast that this correction makes an appreciable difference and gives rise to the magnetic field.

The concept of a field is really an abstract one, because it only describes how one charged particle would behave in the presence of another *if* the second charged particle were there. The interaction does not exist when only one charged particle is present; hence, we could say that the field created by this particle does not exist, but it is not necessary to do so. The field is a very useful concept and helps greatly in understanding electromagnetic phenomena.

If you have difficulty in comprehending action at a distance when, for example, one magnet repels another one, remember that all forces in nature are action at a distance. A hammer does not 'really' hit a nail, the surface atoms get close enough to experience exchange of momentum through a strongly repulsive electric force, this is action at a distance.

If a charge in the form of a current travelling along a wire (Fig. 4.1(*a*)) was moving in a magnetic field, it would experience a force at right angles to both the magnetic field and the direction of current flow. This fact is usually learned as 'Fleming's right-hand rule' and is the principle behind most electrical motors.

To conserve energy, the magnet that creates the magnetic field must feel an equal force pushing it in the opposite direction to the wire. Therefore, there is a magnetic field created by the wire which acts against the 'current' within the magnet. This current is the motion of the electrons in the solid.

We can now see that, if the wire has its own magnetic field, then we can create a magnet by coiling the wire in the form of a cylinder (a solenoid), as shown in Fig. 4.1(*b*), so that all the fields from each segment of wire add up at the centre of the solenoid to create one large magnetic field.

---

†It is interesting to see how classical mechanics is valid only in the narrow area of our direct experience, sandwiched between the mechanics of the very big (relativity) and the very small (quantum mechanics).

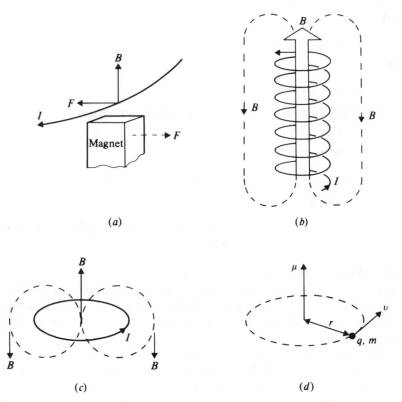

Figure 4.1 Magnetic forces (*F*). (*a*) Wire carrying an electrical current (*I*) in a magnetic field (*B*); (*b*) a solenoid; (*c*) a single current loop; (*d*) a single charge travelling around a loop (charge *q*; mass *m*; velocity v; radius of loop *r*; magnetic moment μ).

If we take one loop of the solenoid, a circle of wire with a current circulating around it, there is a resultant magnetic field perpendicular to the plane of the loop at the centre (Fig. 4.1(*c*)).

This current loop, with its resultant magnetic field, will resist a change in orientation in an externally applied magnetic field in much the same way that a gyroscope resists movement in a gravitational field. There is a form of magnetic inertia called the magnetic moment, which can be related to the mechanical inertia (the angular momentum). The magnetic moment is the current flowing round the loop times the area of the loop.

If we consider a single charge *q*, travelling around a loop of radius *r* at velocity *v* (Fig. 4.1(*d*)), then current is the charge times the number of times it goes round the loop in unit time, i.e.,

Current $= qv/2\pi r$,

and

Area of the loop $= \pi r^2$.

Hence,

Magnetic moment, $\mu = qvr/2$.

The angular momentum p_θ is just the simple momentum ($p = mv$) times the radius r; thus,

$p_\theta = mvr$

(where m is the mass of the charge carrier) and, therefore,

$\mu = (q/2m)p_\theta$.

Hence, we have found the classical relationship between the magnetic moment and the angular momentum.

4.3 Atomic magnets

An electron carries unit charge and the charge distribution is constantly in motion in an atomic stationary state. In fact, the electron has its own magnetic moment, and all the non-spherically symmetric stationary states (p, d, and f) have a magnetic moment.

The relationship between magnetic moment and angular momentum for a stationary state, by a fluke of nature, looks remarkably like that derived classically for a charge circulating around a wire, except that:

1. the electron charge is negative, so we have a minus sign;
2. the expression describes the *component* of the magnetic moment resolved along the direction of the applied field (traditionally called the z direction), and so

$\mu_z = -(q/2m)p_{\theta z}$.

Let us consider the allowed values of $p_{\theta z}$. We know, from de Broglie, that the simple momentum is related to wavelength ($p = h/\lambda$) and that any bound system has a set series of allowed values of wavelength: the quantum states. It should not be difficult to appreciate then that $p_{\theta z}$ is also related to wavelength and will, therefore, have a series of allowed states. In fact, for the angular momentum of a stationary state,

$p_{\theta z} = (h/2\pi)m_l$,†

where m_l is the electron stationary-state (sometimes misleadingly called the 'orbital') magnetic quantum number.

†We can compare this expression with the expression for a violin-string mode derived in Sec. 1.3: $\lambda = 2l/n$. By using the de Broglie expression, $p = h/\lambda$, we find that $p = (h/2l)n$ for a violin-string mode. One can see that this is similar to the angular momentum expression.

We can now write an expression relating the magnetic moment of a stationary state to its magnetic quantum number:

$$\mu_z = -(qh/4\pi m)m_l$$

or

$$\mu_z = -\mu_\beta m_l,$$

where μ_β is a constant and is called the Bohr magneton.

4.4 The quantum numbers

We need now to specify the values of the four types of quantum number.

The scheme for the organization of the stationary states is illustrated in Table 4.1 for the first three values of n. The rules that we apply are quite simple.

1. n is the principal quantum number, specifying the overall mode of the stationary state (analogous to violin-string state numbers).
2. l specifies the type of stationary state (the 'orbital' quantum number) and $l = n - 1, n - 2, \ldots, 0$. If we compare this with Table 1.5, when we first named the electrons, we see that, for s states, $l = 0$; for p states, $l = 1$; for d states, $l = 2$; and for f states, $l = 3$.
3. m_l is the stationary state ('orbital') magnetic quantum number. m_l forms a set of quantum numbers given by the series $l, l - 1, l - 2, \ldots, 0, \ldots, -(l - 2)$, $-(l - 1), -l$. Thus, for s states $(l = 0)$, $m_l = 0$; for p states $(l = 1)$, $m_l = 1, 0$, -1; for d states $(l = 2)$, $m_l = 2, 1, 0, -1, -2$; and for f states $(l = 3)$, $m_l = 3, 2$, $1, 0, -1, -2, -3$.
4. m_s is the electron's own magnetic quantum number and has only two values: $+\frac{1}{2}$ and $-\frac{1}{2}$ (spin up and spin down).

We now see the full energy-level scheme in terms of quantum numbers. For example, for $n = 3$ we have $l = 0$ (s states), $l = 1$ (p states), and $l = 2$ (d states). For the d states we have five values of m_l: $2, 1, 0, -1, -2$. For each of these five values there are two values of m_s, giving us the ten $3d$-type quantum states.

4.5 Quantized magnetic moments

We can now predict that for a p-type orbital $(l = 1)$ $m_l = 1, 0$, or -1 and therefore we can have three values of magnetic moment μ_z: $-\mu_\beta$, 0, or $+\mu_\beta$. This is illustrated in Fig. 4.2(b), where the three allowed directions of the magnetic moment μ are shown,† as well as the projection of this value on the z direction μ_z. In the classical case, all values of μ_z between $+\mu$ and $-\mu$ are allowed.

The quantization of magnetic moment places no restriction on the orientation in the x and y directions; the electron magnet can, therefore, be pictured

---

†The value of μ, the actual magnetic moment, can be worked out statistically as the angular momentum has to be averaged out over all $2l + 1$ quantum states; the result is that $\mu = -\mu_\beta[l(l + 1)]^{1/2}$

Table 4.1

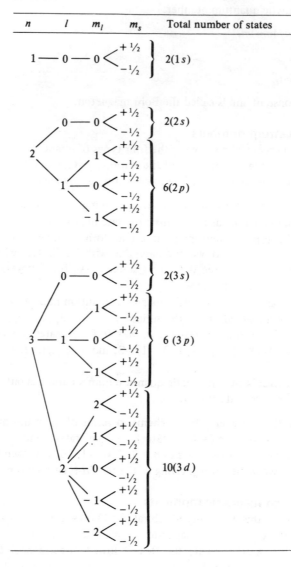

n	l	m_l	m_s	Total number of states

with the magnetic moment precessing around the direction of the magnetic field in the way that a gyroscope precesses around the direction of a gravitational field. We will use this idea of the precession of magnetic moment later, when we briefly consider diamagnetism.

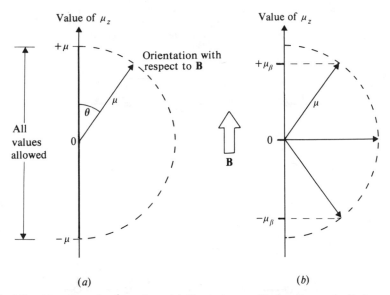

Figure 4.2 The allowed values of μ_z. (*a*) Classical case; (*b*) quantum-mechanical case for $l = 1$.

What happens to the energy-level scheme as a consequence of the quantized magnetic moments?

To answer this question let us consider again the case where $l = 1$ (a p state). The effect of applying a magnetic field is shown schematically in Fig. 4.3. (We are just considering the orbital effects and so we are ignoring for the present the effects of electron spin.) The single p level at energy E_0 splits into three energy levels.

The energy change in a magnetic field, ΔE_{mag}, is just the z component of the magnetic moment times the magnetic field strength B, with a minus sign to denote that there is an energy saving if the magnetic moment is aligned with the field:

$$\Delta E_{mag} = -\mu_z B.$$

For $l = 1$, $\Delta E_{mag} = \mu_\beta B$, 0, or $-\mu_\beta B$.

This spitting of energy levels in a magnetic field can be observed in the spectrum of light emitted by electrons decaying from an excited state. In our example, an electron decaying from $l = 1$ to $l = 0$ levels would emit photons of energy $E = h\nu$ (a spectral line) in the absence of a magnetic field. The line will split into three closely spaced spectral lines when a magnetic field is applied. (The level at $l = 0$, the s state, will not split as $\mu_z = 0$.) The spacing of the lines will be proportional to the magnetic field, reflecting the splitting of the energy levels.

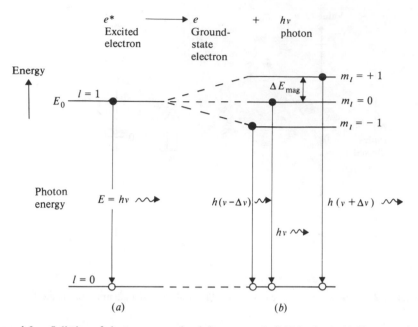

Figure 4.3   Splitting of electron energy levels in a magnetic field for $l = 1$. (*a*) No magnetic field; (*b*) magnetic field **B**.

4.6   Electron spin magnetic moment

The situation for the magnetic moment of the electron, which exists independently of the type of stationary state, is slightly different to that for stationary states.

You will note that the allowed values of the quantum number m_s are $+\frac{1}{2}$ and $-\frac{1}{2}$. In this case,

$$\mu_z = -2\mu_\beta m_s = -\mu_\beta \text{ or } +\mu_\beta.\dagger$$

Thus, in a magnetic field, all energy levels will split into two levels, one higher in energy by $\mu_\beta B$ and the other lower in energy by $\mu_\beta B$, giving a total 'spin splitting' of $2\mu_\beta B$.

The spin splitting must be applied to all energy levels; this means that all the levels shown for 'state splitting' in Fig. 4.3 should really be split into two in a magnetic field. This rather complicates the picture, but transitions can simplify to that shown in Fig. 4.3 if the transitions that are allowed are restricted by occupancy of some states and the rules of quantum mechanics (some transitions are forbidden).

†To aid the memory, one can think of the extra factor of 2 (i.e., $-2\mu_\beta m_s$) as being there to compensate for the fact that the quantum number is not a whole integer.

4.7 Resultant atomic magnetic moments

To this point, we have only considered single states; however, a multi-electron atom has a large collection of states, each with two spin orientations, and the overall magnetic moment of an atom is the resultant of all the component magnetic moments.

It can be seen by the symmetry of Fig. 4.2 that, if all states for a particular value of l (including the pairs of opposite spin) are occupied, there will be no resultant magnetic moment. This occurs because both m_l and m_s are symmetrical about the orientation equivalent to $\mu_z = 0$ (μ perpendicular to the z direction).

Once more, the valence electrons are all-important. They will determine the overall magnetic moment of the atom by the number of states they occupy. The total atomic magnetic moment is a combination of spin and state magnetic moments, given by

$$\mu_z = -g\mu_\beta m_J,$$

where m_J is the total magnetic quantum number and g (called the Landé factor) varies between 2 for pure spin magnetic moments and 1 for pure state magnetic moments.

The idea of the cancellation of magnetic moments can also be applied to collections of atoms. For example, in covalently-bonded solids, the bonding electrons are paired for spin and the bonding states are all occupied, so there will be no resultant magnetic moment for each atom. Therefore, the solid cannot be magnetized. (Magnetization is the alignment of magnetic moments by a magnetic field.)

4.8 The Stern–Gerlach experiment

The quantization of magnetic moment, with only certain allowed values of the component in the direction of the magnetic field, was rather a strange idea to physicists brought up in the classical approach. However, this phenomenon was demonstrated experimentally by Stern and Gerlach in 1922. They produced a beam of silver atoms by evaporating them from an oven into a vacuum, as shown in Fig. 4.4. They could measure the magnetic moment of the silver atoms by passing the beam between the pole tips of a magnet designed to give a very non-uniform magnetic field. This was done by making one pole tip with a very sharp edge whilst the other was flat. The rapidly changing magnetic field had the effect of creating a force which pushed atoms in the same direction as μ_z, the component of the magnetic moment in the direction of the magnetic field. The magnitude of the force is proportional to the value of μ_z. If all values of orientation of magnetic moment were allowed, then the beam of atoms with randomly-oriented magnetic moments would spread out vertically by an equal amount up and down, and would form a strip of deposit on a glass plate that collects the beam, the height of the strip giving the value of magnetic moment.

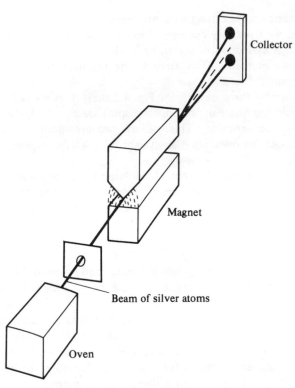

Figure 4.4   The Stern–Gerlach experiment.

In fact, two spots were formed, the silver atoms were split into two distinct beams. One was deflected upwards and the other downwards. Therefore, only two values of μ_z were allowed. It had been proved that magnetic moment was quantized.

4.9   Magnetization of solids

We will now consider what happens when we apply a magnetic field to a large collection of atoms bonded to form a solid.

Every atom will have the same total magnetic moment, with a series of allowed values of the z component μ_z; some of the valence electron energy levels corresponding to a particular value of μ_z will be occupied and some will be vacant. We know that all these levels will have the same energy in the absence of a magnetic field, and, therefore, the same probability of occupation. Both m_l and m_s (and, therefore, m_J) are symmetrical about zero (magnetic moment perpendicular to magnetic field direction); this means that, even if each atom has a resultant magnetic moment, because of the mixing effect of thermal vibrations,

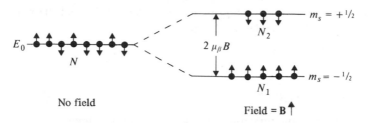

Figure 4.5 Magnetization of a solid with resultant atomic magnetic moment of pure spin. (Arrows indicate direction of μ_z for individual atoms.)

there will be equal numbers in states corresponding to magnetic moments with equal and opposite values of the z component. The total magnetic moment of the solid, M, will be zero.

This argument does not apply to the ferromagnetic elements. However, as there are only three or four of these out of over one hundred elements, these can be considered as important freaks and as such will be dealt with later.

The total magnetic moment of the solid, M, is called the magnetization; we can think of it as the average magnetic moment of the atoms $\langle \mu_z \rangle_{av}$ times N, the number of atoms per unit volume: $M = N \langle \mu_z \rangle_{av}$. In the absence of a magnetic field, $\langle \mu_z \rangle_{av} = 0$.

If we place the solid in a magnetic field, the energy levels will split, so that the energy for electrons of an atom with a magnetic moment aligned with the field is less than that for an atom with magnetic moment aligned against the field.

If we take, for example, a solid in which the resultant atomic magnetic moment is purely due to electron spin, then the energy level at E_0 in the absence of a magnetic field will split into two levels at energies $E_0 + \mu_\beta B$ and $E_0 - \mu_\beta B$, as shown in Fig. 4.5.

The majority of the atoms will occupy the lower level, and the solid as a whole will have a magnetic moment; it has been magnetized.

Because of the statistical variation in possible energies due to thermal vibration (in this case only two energies are possible), some atoms will occupy the higher energy level, with μ_z opposed to the magnetic field.

We can calculate† quite easily the relationship between M, B, and T (temperature) using just one rule: that the probability of an electron occupying a level shifted by an energy ΔE_{mag} is proportional to $\exp(-\Delta E_{mag}/kT)$,‡ where k is Boltzmann's constant.

---

†This is the only mathematical proof in this book and no apologies are made for including it here because it is elegant, simple, and it is sensible to see one example of how mathematics is used for a quantitative description of the properties of solids.
‡Remember we used a similar expression when determining the distribution of free electrons and the Fermi energy in a solid.

If the number of atoms occupying the lower level is N_1 and the number of atoms occupying the upper level is N_2, then

$$N_1 = C \exp(\mu_\beta/kT)$$

and

$$N_2 = C \exp(-\mu_\beta B/kT),$$

where $N_1 + N_2 = N$, the total number of atoms per unit volume.

Using the above three expressions, we can find the value of the constant C:

$$C = N/[\exp(\mu_\beta B/kT) + \exp(-\mu_\beta B/kT)].$$

The resultant excess of atoms aligned with the magnetic field as a fraction of the total number is

$$\frac{N_1 - N_2}{N} = \left\{ \frac{N \exp(\mu_\beta B/kT)}{[\exp(\mu_\beta B/kT) + \exp(-\mu_\beta B/kT)]} - \frac{N \exp(-\mu_\beta B/kT)}{[\exp(\mu_\beta B/kT) + \exp(-\mu_\beta B/kT)]} \right\} \Big/ N$$

$$= \left(\frac{\exp(\mu_\beta B/kT) - \exp(-\mu_\beta B/kT)}{\exp(\mu_\beta B/kT) + \exp(-\mu_\beta B/kT)} \right)$$

$$= \tanh(\mu_\beta B/kT).$$

As we have pure electron spin, we know from Sec. 4.6 that, for each atom aligned with the field, $\mu_z = +\mu_\beta$; hence, the average magnetic moment per atom in the solid is given by

$$\langle \mu_z \rangle_{av} = \mu_\beta \tanh(\mu_\beta B/kT)$$

and the magnetization of the solid is given by

$$M = N\mu_\beta \tanh(\mu_\beta B/kT).$$

The relationship between M and $\mu_\beta B/kT$ is shown in Fig. 4.6.

It is known that, as x approaches infinity, $\tanh x$ approaches 1, and we can see here that for very high fields and very low temperatures all the atoms occupy the lower energy level (all the magnetic moments are aligned with the field) and $M = N\mu_\beta$. The magnetization is saturated.

At normal temperatures and fields that can be achieved easily in the laboratory, $\mu_\beta B/kT$ is very small (0.02 for room temperature and $B = 1$ Wb m^{-2}) and $\tanh x$ approaches x as x approaches 0. Hence,

$$M \simeq (N\mu_\beta^2/kT)B. \tag{4.1}$$

Thus, magnetization is proportional to the magnetic field (at constant temperature).

The phenonmenon of the alignment of the atomic magnets by a magnetic field is the most common form of magnetism for atoms with a net magnetic moment, and so it is rather strange that this is called paramagnetism.

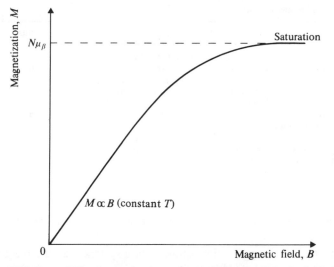

Figure 4.6 Variation of the magnetization of a solid with magnetic field (assuming constant temperature).

4.10 Magnetic susceptibility

A constant that tells us how much a solid can be magnetized (how susceptible it is to magnetization) is denoted by the symbol χ and is called the magnetic suceptibility.

Susceptibility is a practical unit which can be measured in the laboratory by measuring the change in magnetic field when the solid of interest is placed at the centre of a solenoid. For this reason, the susceptibility is defined in terms of the engineer's magnetic field H:

$$M = \chi H \tag{4.2}$$

(assuming that $\mu_\beta B \ll kT$, so that magnetization is proportional to magnetic field).

H is defined in terms of the electric current one would supply to the solenoid. Therefore, H is defined in terms of ampere turns per metre of solenoid. The units of H are amperes per metre.

In Sec. 4.2, we defined magnetic moment as the current flowing round a loop times the area of the loop, so magnetic moment has the dimensions of amperes $\times$ metre^2. As magnetization is magnetic moment per unit volume, magnetization will have the same units as H (amperes per metre). The susceptibility is, therefore, a dimensionless number.

Magnetic units and their definitions are shown in Table 4.2. The magnetic

Table 4.2   The magnetic units

Quantity	Symbol	Definition	Units
Magnetic moment	μ	Current x area	Ampere x metre^2
Magnetization	M	Magnetic moment/volume	Ampere/metre
Magnetic field (imposed)	H	Ampere turns/ circumference	Ampere/metre
Susceptibility	χ	$M = \chi H$	Dimensionless
Magnetic field (induced)	B	$F = q\mathbf{v} \times \mathbf{B}$	Weber/metre^2 = tesla†
Permeability of free space	μ_0	$B = \mu_0 \mu_r H$	Weber/metre ampere
Relative permeability of a medium	μ_r	$B = \mu_0 \mu_r H$	Dimensionless

† 1 Wb m^{-2} = 1 Te = 10^4 gauss in e.m.u.

field **B** is defined in terms of the force **F** exerted on charge carriers of charge q and velocity **v**.†

Thus, B is the magnetic field induced *in* a system, i.e., the magnetic effect. While H is the magnetic field intensity imposed *on* the system from outside.

The constant of proportionality between B and H is called the magnetic permeability and

$$B = \mu_0 \mu_r H, \tag{4.3}$$

where μ_0 is the permeability of free space and incorporates the change of units between B and H; μ_r is the permeability of any medium within which B is acting relative to that of free space ($\mu_r = 1$ for free space) and is related to the susceptibility.

The relationship between χ and μ_r can be determined if we consider that the induced magnetic field (B) is a combination of the externally applied field (H) and the internal magnetization (M). Thus,

$$B = \mu_0 (H + M) \tag{4.4}$$

(μ_0 to allow for the change of units). Therefore,

$$\mu_0 (H + M) = \mu_0 \mu_r H$$

(from Eq. (4.3)) and

$$H + \chi H = \mu_r H$$

(from Eq. (4.2)). Hence,

$$\chi = \mu_r - 1. \tag{4.5}$$

When χ is very small, μ_r is close to 1, when χ is large $\mu_r \simeq \chi$.

†A vector equation with vector multiplication has to be used because **F**, **v**, and **B** act in different directions – remember Fleming's right-hand rule.

The value of χ is closely related to the resultant atomic magnetic moment. This can be seen in the relationship derived at the end of Sec. 4.10 (Eq. (4.1)), where magnetization is effectively proportional to the square of the resultant magnetic moment (μ_β) times the applied field. This implies that, for small values of χ, susceptibility is proportional to the square of the atomic magnetic moment. The value of χ for the 83 elements for which χ is known is shown in Table 4.3. In this table, the elements are arranged so that the total number of

Table 4.3

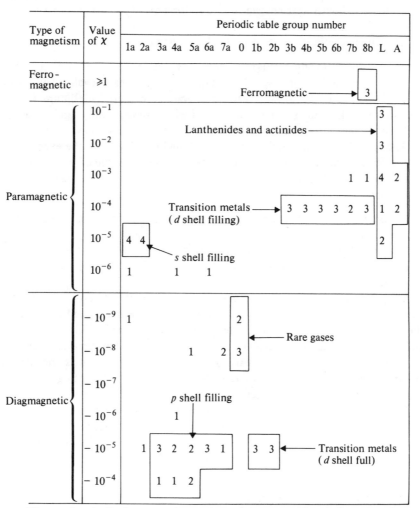

Type of magnetism	Value of χ	Periodic table group number																
		1a	2a	3a	4a	5a	6a	7a	0	1b	2b	3b	4b	5b	6b	7b	8b	L	A
Ferro-magnetic	≥ 1																3 (Ferromagnetic →)		
Paramagnetic	10^{-1}																	3	
	10^{-2}					Lanthenides and actinides →												3	
	10^{-3}															1	1	4	2
	10^{-4}				Transition metals (d shell filling) →							3	3	3	3	2	3	1	2
	10^{-5}	4	4				(s shell filling)											2	
	10^{-6}	1			1		1												
Diamagnetic	-10^{-9}	1							2 (Rare gases)										
	-10^{-8}					1		2	3										
	-10^{-7}																		
	-10^{-6}			1 (p shell filling)															
	-10^{-5}	1	3	2	2	3	1			3	3								
	-10^{-4}		1	1	2														

(Transition metals, d shell full)

The numbers in the boxes are the numbers of elements with a similar value of χ.

elements in each group of the periodic table with a value of χ of the same order of magnitude are placed in the same box. For example, all elements for which the value of χ falls between 0.05 and 0.005 would be placed in the 10^{-2} boxes.

It is very clear from Table 4.2 that the valence-electron configuration determines the value of χ.

The elements with the highest susceptibility are the three Group 8 ferromagnetic transition metals. These can retain some magnetization at zero field and so, in terms of Eq. (4.2), they will have infinite susceptibility. Practically, they can be given a value of susceptibility, as we will explain when we come to discuss ferromagnetism, but for now it is sufficient to say that $\chi \geqslant 1$.

The paramagnetic elements fall into three categories.

1. The lanthenides (rare earths) and actinides, where the f shell is filling, have a high susceptibility (although there is quite a large spread in values) with three of the rare earths having a very high value. In fact, gadolinium (atomic number 64), with $\chi = 0.48$, has the highest susceptibility and is weakly ferromagnetic at just below room temperature.
2. The early transition metals, where the d shell is filling (with the exception of the ferromagnetic elements), which nearly all have a value of χ around 10^{-4}.
3. Groups 1a and 2a, where the s shell is filling, which nearly all have a value of χ around 10^{-5}.

This grouping tells us that the more complex the valence-electron configuration, the greater will be the resultant atomic magnetic moment.

The model used to explain magnetization of solids will not predict negative values of χ, but we can see that they exist. This is the phenomenon of diamagnetism.

The effect is small for the rare gases, which have almost zero χ due to their closed-shell structure; however, both the groups where the p shell is filling and the transition metals where the d shell is full and the s shell is filling (Groups 1b and 2b) have values of χ around -10^{-5}. The most diamagnetic element is bismuth (atomic number 83), with $\chi = -1.6 \times 10^{-4}$. Bismuth is the heaviest element, apart from four actinides, for which the value of χ is known. Diamagnetism will be briefly discussed in Sec. 4.12.

It is difficult to find a use for the paramagnetic effect, but whenever a magnetic field penetrates a solid it is modified by the magnetization of that solid. However, we can give an example of the use of paramagnetism in the research laboratory to cool samples to very low temperatures.

4.11 Cooling by adiabatic demagnetization
Paramagnetism can be used to achieve temperatures very close to absolute zero.

A salt containing a rare earth element of very high susceptibility is cooled by maintaining it in contact with liquid helium to about one degree above absolute zero ($-272\,^\circ$C or 1 K).

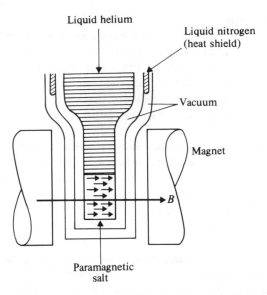

Liquid helium

Liquid nitrogen
(heat shield)

Vacuum

Magnet

B

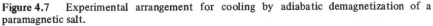

Paramagnetic
salt

Figure 4.7 Experimental arrangement for cooling by adiabatic demagnetization of a paramagnetic salt.

The salt is then magnetized using a large magnet, which is outside the vacuum system and heat shields. At this temperature, all the magnetic moments will align with the field and we have saturation magnetization. This is illustrated in Fig. 4.7.

The liquid helium is then pumped away, leaving the fully magnetized salt thermally isolated in a vacuum.

The magnetic field is then slowly reduced to zero. As the field is reduced, the atomic magnets flip over to assume random orientation with $\langle \mu_z \rangle_{av} = 0$; however, as some field remains, this 'flipping' needs energy, which is taken from the thermal vibrations.

Randomness in magnetic moment orientation is bought at the expense of reduction in thermal energy. The temperature of the salt plunges down to about one thousandth of a degree above absolute zero.

This is not the end of the story, because the nucleus of an atom also has a magnetic moment.† A salt with a high nuclear magnetic moment is cooled by the electron paramagnetic salt and can then be used for a second stage of cooling by demagnetization to achieve temperatures only one millionth of a degree away from absolute zero.

---

†Nuclear paramagnetism is very much smaller than electron paramagnetism, because $\mu_{\beta(nuclear)} = qh/4\pi m_p$, where m_p is the proton mass (which is two thousand times greater than the electron mass).

4.12   Diamagnetism

It seems strange that a material should have a negative susceptibility. If the atoms had a resultant magnetic moment, this would imply that they prefer to occupy the higher energy level opposed to the magnetic field. This cannot occur, and, in fact, we only observe the diamagnetic effect when the atomic magnetic moment is zero. We must point out, however, that the diamagnetic effect occurs for *all* atoms, but is masked by the larger positive susceptibilities of paramagnetism and ferromagnetism.

The diamagnetic effect arises from electromagnetic induction. Think of our classical current loop, first considered in Sec. 4.2. Lenz's law tells us that, if we change the magnetic field by ΔB, then the current passing around the loop will change in such a way as to oppose the change in field. One can visualize this change in current as a change in the velocity of the charge carrier by Δv.

The magnetic moment will change by an amount $\Delta \mu$ proportional to Δv.

If we now consider the electron wave, $E = h\nu$, where ν is the frequency of the wave. The frequency, in the classical case, is directly proportional to the velocity of the charge carrier; this is also true for electron waves.

In a magnetic field, $\Delta E_{mag} = -\mu_z \Delta B$ (Sec. 4.5), i.e., $\Delta E \propto -\Delta B$, so we can lump the proportionalities together to say that

$$\Delta \mu \propto \Delta v \propto \Delta \nu \propto \Delta E \propto -\Delta B,$$

i.e.,

$$\Delta \mu \propto -\Delta B.$$

In other words, the change in magnetic moment opposes, and is proportional to, the change in magnetic field. We have a negative susceptibility.

In the classical model, the magnetic moment is proportional to loop area, and it is interesting to note that the largest non-paramagnetic element for which there is data available, bismuth, has the biggest negative susceptibility. Of course, we cannot put an exact loop radius into the equation for the diamagnetic change in the magnetic moment of an atom. We have to use a mean radius incorporating the spread of electron wave functions.

4.13   Ferromagnetism

A couple of hints about ferromagnetism have been made earlier in this book. In Sec. 3.5, we saw that iron, with a valence configuration of $3d^6 4s^2$, has a band structure (Fig. 3.10(*b*)) with complete overlap of the $3d$ band by the $4s$ band, and the Fermi level is such that the $3d$ band is not completely full. It was stated that the $3d$ band is the key to the property of ferromagnetism possessed by iron and the two neighbouring Group 8 metals, cobalt ($3d^7 4s^2$) and nickel ($3d^8 4s^2$).

We will now examine the $3d$ band in more detail and describe a crude model that will explain the unique property of ferromagnets: magnetization at zero field (infinite susceptibility). We will then see how the presence of magnetic

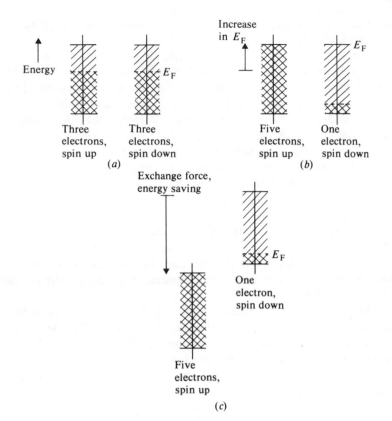

Figure 4.8 The 3*d* band for iron, divided into spin up and spin down components. (*a*) Three spin up, three spin down; (*b*) five spin up, one spin down; (*c*) the same as (*b*) plus the exchange force.

domains means that real ferromagnets need to be magnetized and, therefore, have a large but finite susceptibility.

In Fig. 4.8, we show the 3*d* band for iron with the band separated into two parts: one for 'spin up' electrons and one for 'spin down' electrons. If we assume that the 3*d* band contains the six 3*d* electrons,† then we would expect that we would have three spin up and three spin down, as shown in Fig. 4.8(*a*). The real situation is that the spin up band is full, with five electrons, leaving just one electron in the spin down band. It would appear that this configuration is not

---

†This is the first of several simplifying assumptions which are not entirely true. In this case, the valence electrons are shared between the overlapping 4*s* and 3*d* bands, so they cannot be exactly allocated to their own band. However, although the bands overlap in energy, they do not overlap in momentum, so the assumption is not as bad as all that.

112

energetically favourable, as it would mean an increase in the Fermi level, as shown in Fig. 4.8(*b*).

In fact, the configuration with a full spin up band is stable, i.e., there is an energy saving, as shown in Fig. 4.8(*c*). This energy saving only occurs *if the spin magnetic moments of neighbouring atoms have the same alignment*. The energy saving is due to electrostatic forces called exchange forces, which act rather like the Pauli exclusion principle. These forces influence the spin orientation of neighbouring atoms through the overlap of the electron wave functions.

The wave functions can only overlap in certain restricted ways and, for the ferromagnetic elements, the overlap works in a way similar to that shown in Fig. 4.9,† where two wave functions of similar spin look like an *s* wave function where they overlap, providing a site where an *s* electron can sit.

It is only for iron, cobalt, and nickel that interatomic separation (atom size) is small enough for the overlap of wave functions to result in large positive exchange forces. For other Group 8 metals of similar valence-electron configuration, the atomic size is larger and the exchange forces too weak to achieve correlation of the spins of neighbouring atoms against the randomizing effect of thermal vibrations.

To help with the understanding of the way this overlap results in an overall magnetization of the solid, Fig. 4.10 shows, diagrammatically, the electron spin distribution in iron.‡ Iron has five 'spin up' 3*d* electrons and one 'spin down'. The 'spin down' 3*d* electron will cancel out the effect of one of the 'spin up' electrons, leaving four 'spin up' 3*d* electrons to interact with neighbouring atoms. Each one of the four 3*d* 'spin up' electrons will overlap, via a 4*s* 'spin

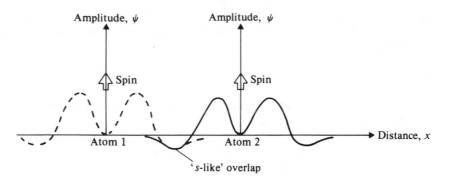

Figure 4.9   Overlap of electron stationary states from neighbouring atoms.

---

†The second of the simplifying assumptions, the manner of the overlap is more complicated than this.
‡The third of our simplifications. Iron is b.c.c. and, therefore, has eight nearest neighbours; the effect is shared between all of them. Also, as we have a 3*d band*, the 3*d* electrons are not confined to the atom but are almost free; therefore, they must be able to interact directly with one another.

3 d electrons: ⇧ spin up, ⇩ spin down
4 s electrons: ↑ spin up, ↓ spin down

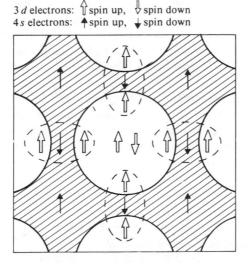

Figure 4.10   Schematic representation of the valence-electron spin distribution in iron.

down' electron, with the 3d 'spin up' electron of a neighbouring atom in such a way that the resultant magnetic moments of the two atoms are parallel. Thus, with our simple model, we predict saturation magnetization with zero magnetic field. (The magnetic moment of the 4s 'spin down' electron that takes part in the exchange overlap is cancelled out by a 4s 'spin up' electron which does not take part in the exchange.)

The spin magnetic moment is μ_β per electron (Sec. 4.6), so the resultant magnetic moment per atom in our simple model for iron is $4\mu_\beta$ and the magnetization (for zero field remember) is $M = M_{sat} = N \times 4\mu_\beta$.

Cobalt has seven 3d electrons and so the resultant electron magnetic moment per atom is $3\mu_\beta$; for nickel, with one more electron, it is $2\mu_\beta$. The exchange forces work for these two elements in a similar way to those for iron, so that the spin magnetic moments of all the atoms in the solid point in the same direction.

4.14   Antiferromagnetism

Manganese, the element just before iron, has a valence electron configuration of $3d^5 4s^2$ and, like the ferromagnetic elements, the 'spin up' band is full, giving a theoretical magnetic moment per atom of $5\mu_\beta$. In spite of this, a crystal of manganese exhibits no overall magnetization.

It so happens that the exchange forces oscillate both in magnitude and sign as the interatomic separation changes. For manganese, the exchange forces are large, but they work in the opposite way to those for the ferromagnetic elements. The wave functions overlap in such a way that the resultant atomic magnetic moments of neighbouring atoms point in opposite directions. This alternation in direction means that the overall magnetic moment of solid manganese is zero. This phenomenon is called antiferromagnetism. Although the ordering in spins does not create a magnetic field, it affects properties such as the electron specific heat, which are affected by changes in the degree of order in a solid. There is a sharp rise in specific heat when the spin order is destroyed at high temperatures.

The interatomic separation for manganese can be altered by materials engineering. This is done by alloying manganese with the non-ferromagnetic elements from the same row of the periodic table: aluminium and copper. By this method, the manganese—manganese separation can be increased slightly, so that the exchange forces change sign, but are still large. The alloy so formed is ferromagnetic.

The same effect can be produced by incorporating manganese in a ceramic metal oxide crystal, creating an electrically insulating ferromagnetic material. This type of ferromagnet is called a ferrite and we will describe the crystal structure of the simplest of the ferrites below.

4.15 Ceramic magnets (ferrites)

Magnets are often used at high frequencies in electrical circuits, but the electrical currents induced by the changing field inside a ferromagnetic metal cause problems due to power loss and heating of the magnet. One way around this problem is to limit the current paths by building the magnet core in the form of many thin laminations (as is done for many transformers). A much more elegant solution, however, is to use an electrically insulating magnet: a ferrite.

The ferrites are usually an oxide of a ferromagnetic (or antiferromagnetic) metal and many types have been produced. We will just describe one: the naturally occurring oxide of iron, called magnetite, Fe_3O_4.

Part of the unit cell of magnetite is shown in Fig. 4.11. The whole unit cell is made up of eight cells similar to that shown, and the structure is called the inverse spinel structure. The bonding is mainly ionic, with two triply-charged iron ions (Fe^{3+}) one doubly-charged iron ion (Fe^{2+}) and four oxygen ions (O^{2-}), i.e.,

$$Fe_3O_4 \equiv Fe^{2+}Fe_2^{3+}O_4^{2-}.$$

In our section of the unit cell, the oxygen ions are arranged on a face-centred cubic lattice with one Fe^{3+} ion at a site of the type $\frac{1}{4} \frac{1}{4} \frac{1}{4}$ (called a tetrahedral site) and one at a site halfway along one edge of the cube, i.e., of the type $\frac{1}{2}$ 0 0 (called an octahedral site). The one Fe^{2+} is at another octahedral site.

The Fe^{3+} ions are missing both $4s$ electrons and one $3d$ electron and so have a resultant magnetic moment of $5\mu_\beta$. However, the exchange forces act on the Fe^{3+} ions to create alternation of spin direction (antiferromagnetism), so cancelling out their overall magnetic effect.

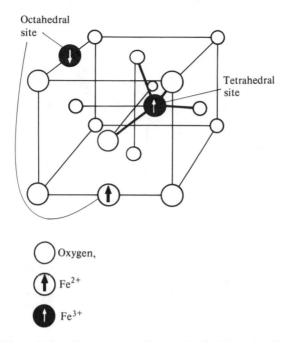

Octahedral site

Tetrahedral site

◯ Oxygen,

Fe^{2+}

Fe^{3+}

Figure 4.11   The structure of magnetite ($\frac{1}{8}$ of the unit cell).

The ferrimagnetism (magnetism of ferrites) arises from the one Fe^{2+} ion which has lost the two $4s$ electrons but has the same $3d$ electron configuration and, therefore, the same magnetic moment as neutral iron, $4\mu_\beta$. Here, the exchange forces act as they do in the ferromagnets and we can have an overall magnetization at zero field.

Most ferrites used in electronics are synthetic, but many are based on magnetite, with the Fe^{2+} ions replaced by manganese, nickel, or cobalt. Non-magnetic ions such as zinc are used to replace some of the Fe^{3+} ions, altering the sign of the exchange forces of those remaining so that their magnetic moments do not cancel out.

Ferrites of this type, or with radically different crystal structures but working on much the same principles as those described, are used for such diverse applications as the production of permanent magnets for closing refrigerator doors, low-loss ferrites for use at microwave frequencies, and 'square loop' magnets for computer memories.

4.16   The effect of temperature on ferromagnets

We have already described some of the effects of thermal vibrations, such as the spreading of electron energies close to the Fermi level to allow conduction in semiconductors. Of more relevance here is the randomization of the magnetic

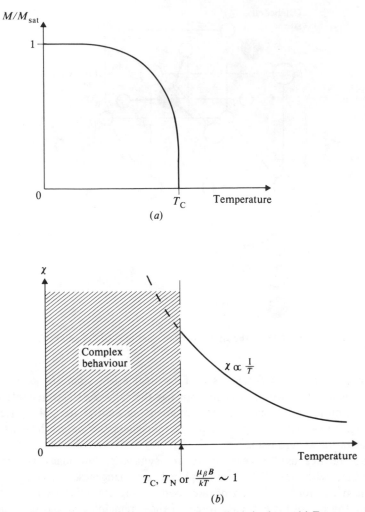

Figure 4.12 The effect of temperature on magnetic behaviour. (a) Ferromagnetism; (b) susceptibility.

moment of paramagnetic solids to allow occupation of the higher energy level when splitting of energy levels occurs in a magnetic field.

Thermal vibrations also attempt to randomize the direction of magnetic moment in ferromagnets, and so work in opposition to the exchange forces. This reduces the spontaneous magnetization (magnetization at zero field) below that

Table 4.4   Data for the ferromagnetic elements

Property	Element			
	Fe	Co	Ni	Gd
Curie temperature T_C (°C)	770	1131	358	16
$\langle \mu_z \rangle_{av}$, simple theory	$4\,\mu_\beta$	$3\,\mu_\beta$	$2\,\mu_\beta$	?
$\langle \mu_z \rangle_{av}$, measured	$2.2\,\mu_\beta$	$1.7\,\mu_\beta$	$0.6\,\mu_\beta$	$7.1\,\mu_\beta$

of saturation magnetization ($4N\mu_\beta$ in our simple model for iron). The magnetization varies with temperature in the way shown in Fig 4.12(a).

When the thermal energy is larger than the energy saving due to the exchange forces, the correlation of magnetic moments breaks down very rapidly and completely. The spontaneous magnetization becomes zero and the solid becomes paramagnetic.

The temperature at which this happens is called the Curie temperature (T_C). The value of T_C for the four ferromagnetic elements (gadolinium is included here) is shown in Table 4.4. Cobalt retains its ferromagnetic properties to very high temperatures, while the rare earth gadolinium becomes paramagnetic at about room temperature.

Also shown in Table 4.4 is the average magnetic moment per atom derived from our simple model compared with the actual values that have been measured. We can see that we have overestimated the value of $\langle \mu_z \rangle_{av}$; this can be taken as evidence of the band overlap between $4s$ and $3d$ bands, with the resultant sharing of electrons between the two bands.

A transition from correlation of spins to random paramagnetic behaviour occurs for antiferromagnets; the temperature at which this happens is called the Néel temperature (T_N).

By analogy, we can see that the ease with which paramagnetic spins can be aligned by external magnetic fields must diminish with increasing temperature. Comparison of Eq. (4.1) with Eq. (4.2) makes it not unreasonable to assume that susceptibility is inversely proportional to temperature.

We summarize the effect of temperature on paramagnetic, ferromagnetic, and antiferromagnetic solids in Fig. 4.12(b). Above a certain transition temperature (T_C, T_N, or $\mu_\beta B/kT \sim 1$), the solid is paramagnetic and $\chi \propto 1/T$. Below that temperature, we have complex behaviour, the material becomes ferromagnetic, antiferromagnetic, or − for paramagnetic solids (where the transition is much more gradual) − we approach saturation magnetization. As a result, M is no longer proportional to B, and χ becomes a function of H.

4.17 Magnetic domains

We will now examine what happens in real ferromagnetic solids. Our simple model predicts complete correlation of spins throughout the crystal structure,

118

resulting in saturation magnetization with zero external magnetic field. But experience tells us that not every piece of iron is a strong magnet. In fact, pure iron, with no defects in the crystal structure, almost invariably has no magnetization at all in the absence of a magnetic field. The reason for this is that, although there is an energy saving through exchange forces by complete correlation of spins, a large amount of magnetic energy is lost in maintaining the external flux between north and south poles of the magnet. This energy loss is reduced by the creation of magnetic domains, as illustrated in Fig. 4.13. The magnetic moments rearrange to divide the crystal into regions that are each fully saturated, but arranged geometrically to minimize the external magnetic flux.

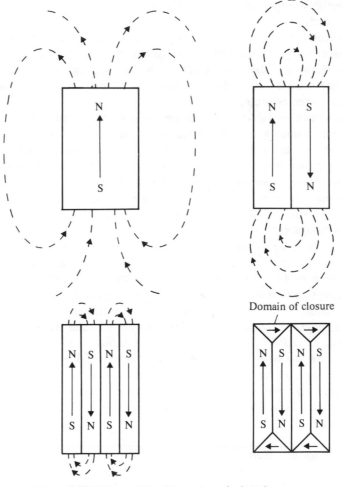

Figure 4.13   Formation of ferromagnetic domains.

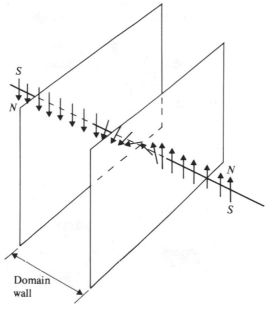

Figure 4.14   A domain wall.

Domains of closure often form which virtually eliminate all external flux by closing the magnetic flux circuit within the solid. This is analogous to putting the 'keep' across the end of a permanent magnet to retain its strength.

The domains are separated by a region of changing magnetic moment, as illustrated in Fig. 4.14. This region, which is called a domain wall, is quite large on an atomic scale (typically 300 lattice constants). Energy is required to create a domain wall, because this has to be done by opposing the exchange forces to a certain extent, although the change in spin direction is gradual. The final size and shape of domains is determined by the balance between the energy saved in reducing the external flux and the energy required to create domain walls.

4.18   Magnetization of ferromagnets

The magnetization of paramagnets is by alignment of atomic magnetic moments with the applied magnetic field. The magnetization of ferromagnets is by quite a different mechanism, that of domain growth induced by an applied magnetic field.

Let us take a sample of iron with a completely equilibriated domain structure (and, therefore, zero magnetization) and examine what happens to the domain structure when we apply an external magnetic field.

At low fields, the domains are usually free to change their shape. Therefore, the effect of an external magnetic field is that those domains where the direction

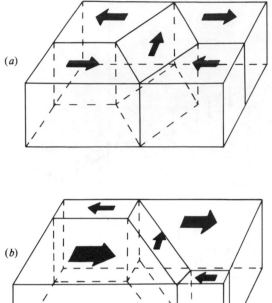

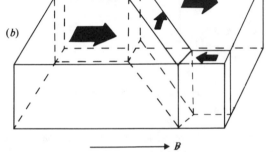

Figure 4.15 Domain growth. (*a*) Equilibrium domain structure; (*b*) external magnetic field *B* applied.

of magnetic moment is close to the field direction will grow at the expense of the domains with a magnetic moment opposed to the field. This change of shape, which is illustrated in Fig. 4.15, is reversible at low fields.

We can illustrate the variation of magnetization M with the external magnetic field H by plotting M against H. Conventionally though, this is illustrated by plotting B (the induced magnetic effect) against H (the externally applied field), where

$$B = \mu_0(H + M)$$

(Eq. (4.4)).

The magnetic effect is a combination of the external field and the magnetization. At small values of H, the second term dominates this equation because it only needs a little magnetic field to encourage the domains to grow, with a resultant huge increase in M. So B is almost directly proportional to M, except at high fields. At high fields, where $M = M_{sat}$ and where we no longer have any domains, there is no further change in magnetization. In this case, B becomes proportional to H.

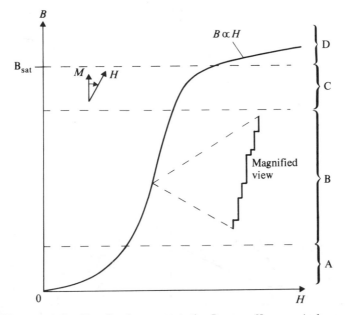

Figure 4.16 Magnetization of a ferromagnet; the B versus H curve. A shows reversible domain growth; B shows irreversible domain growth; C shows magnetization rotation; D shows saturation magnetization.

The relationship between B and H for a typical iron sample is shown in Fig. 4.16. The region of reversible domain growth at low fields is often followed by a region of irreversible domain growth. This arises because domain growth is stopped by imperfections in the crystal, such as impurities, dislocations, and grain boundaries. These act as barriers that domain walls cannot cross until a critical value of field is reached; at such critical field values, there is enough energy for the domain wall to overcome the barrier. Almost instantaneously, the domain expands until it reaches the next barrier. This results in a sharp increase in magnetization and, if the B/H relationship is inspected closely, it will be seen to consist of a series of small steps, each one representing the leap of an individual domain over a barrier in the crystal. This is called the Barkhausen effect.

At sufficiently high fields, all the barriers will be overcome and the whole solid becomes one domain; we have saturation magnetization (assuming we are well below the Curie temperature), but the direction of magnetization is the easy magnetization direction in the crystal structure. There is a further apparent increase in magnetization as the direction of magnetization is rotated out of the crystalline direction to be exactly parallel to H. At fields above this, M does not change and B is proportional to H.

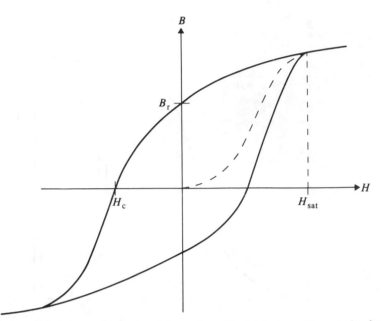

Figure 4.17   The ferromagnet hysteresis loop. B_r, residual induction; H_c, coercive force; H_{sat} magnetic field for saturation magnetization.

4.19   The hysteresis loop

Consider now a solid that has exhibited irreversible domain growth and is at a value of H where it is fully magnetized. When the magnetic field is reduced, the energy saved by creation of domains is not always enough to overcome the barriers to the movement of domain walls in the crystal. The result of this is that, at zero external field, there are still more domains aligned with the original direction of the magnetic field than opposed to it. The solid retains some magnetization. This is characterized by the residual induced magnetic field B_r.

The magnetic field has to be reversed and raised to a value H_c (called the coercive force) in order to push domain walls over the barriers so that we regain zero magnetization.

The solid can be magnetized in the opposite direction if the reversed field is increased. The B/H curve traces out a hysteresis loop which is shown in Fig. 4.17; the area of the loop is a measure of the energy expended during the magnetization cycle from $+H_{sat}$ through $-H_{sat}$ and back to $+H_{sat}$, due to the necessity of pushing domain walls through regions of crystal imperfection.

4.20   Three case histories

The materials engineers can create ferromagnetic and ferrimagnetic materials with a great range of properties by controlling the barriers to domain growth. Three examples of how widely different requirements are met will be described here.

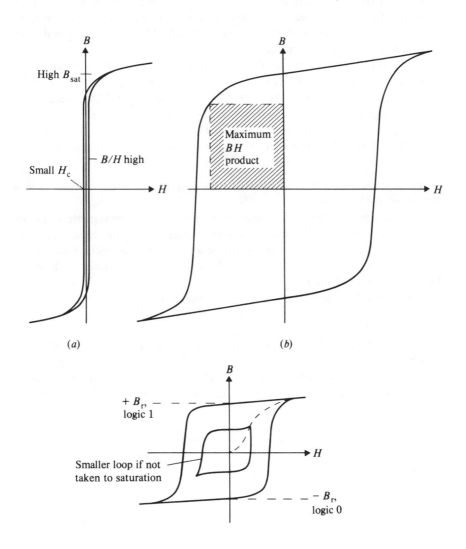

Figure 4.18 Examples of ferromagnetic hysteresis loops. (*a*) The transformer core; (*b*) the permanent magnet; (*c*) the square-loop ferrite.

(a) The transformer core

The requirements for the core of an electrical transformer which forms the magnetic link between two or more electrical circuits are summarized as follows.

1. Very large magnetization for small fields, so that a good linkage is established between the two electrical circuits. This is equivalent to requiring a very large value of relative permeability μ_r and, as we know that $\chi \cong \mu_r$ for $\mu_r \gg 1$, this is identical to a very large value of susceptibility. The permeability is defined as the slope of the B/H curve.
2. Low losses in the core, so that we do not waste energy that should be transferred between the electrical circuits in heating the core. This implies the minimum hysteresis loop area, which can only be achieved by having a very small value of coercive force (H_c) because of requirement 3, below.
3. A large saturation induction B_{sat} (where $B_{sat} = \mu_0(H_{sat} + M_{sat})$), so that the output waveform follows the input waveform at the maximum amplitude required.

The hysteresis curve of a material ideal for use in a transformer core is illustrated in Fig. 4.18(a).

These properties, which are those of a very 'soft' ferromagnetic solid, are achieved by using a metal with as few impurities and crystal imperfections as possible. The metal must be well annealed and handled carefully, because any mechanical abuse, such as bending, will introduce dislocations which act as barriers to domain growth.

Some typical properties of transfer core materials are given in Table 4.5(a). The alloys are all single phase, as a second phase would also act as a barrier to domain growth.

Some of the best alloys are the permalloys, which are nickel–iron alloys. As well as being ideal in terms of the criteria above, they have other important properties: at 70 per cent nickel, the alloy has no preferred direction of magnetization, as the preference for nickel to magnetize in the [111] direction

Table 4.5   Properties of ferromagnetic alloys

Alloy	μ_r	H_c (A m^{-1})	B_{sat} (Wb m^{-2})	B_r (Wb m^{-2})
(a) The transformer core alloys				
4 per cent Si/Fe	6.5×10^3	40	2.0	
Pure Fe	3.5×10^5	0.9	2.2	
Permalloy	1×10^6	0.3	0.8	
(b) Permanent magnet alloys				
Alnico V †		5.4×10^4		1.3
Palatinax		4×10^5		0.6

† Alnico V is 8 per cent Al, 14 per cent Ni, 24 per cent Co, 3 per cent Cu, 51 per cent Fe.

is balanced by the preference of iron to magnetize in the [100] direction; at 80 per cent nickel, the magnetostrictive effects (change of crystal shape with magnetic field) of nickel and iron, which are of opposite sign, are equal in magnitude. Thus, magnetostrictive effects which can lead to losses in the core and transformer 'hum' can be eliminated.

(b) The permanent magnet

The requirements for a permanent magnet are very simple. A very large magnetization must be retained at zero field (B_r) and a very large reverse field must be required before this is lost, implying a large value of coercive force (H_c).

The quality is best assessed by measuring the maximum $B \times H$ product (the area of the largest rectangle) in the second quadrant of the hysteresis curve.

A typical hysteresis curve is shown in Fig. 4.18(b) and the properties of two permanent magnet alloys are given in Table 4.5(b). The values of B_r are comparable with B_{sat} for the transformer alloys, but the value of H_c is up to six orders of magnitude greater for the permanent magnet alloys.

The requirements imply that domain movement and creation of domain walls should be impossible. Reversal of magnetization can then take place only by complete reversal of the direction of magnetization of the domain. This is achieved in the quinternary alloy Alinco V by the precipitation of the magnetic phase in a non-magnetic matrix. The precipitates are roughly rectangular and a few thousand angstroms long by a few hundred angstroms wide. The shape of the precipitates discourages the formation of domain walls. They are very elongated and all parallel. The solidification takes place in a strong magnetic field, which is parallel to the long axis of the precipitates. Each magnetic phase precipitate is a fully saturated single domain. The domain is unable to move as the magnetic phase is embedded in a non-magnetic matrix.

(c) A computer memory

One type of computer memory takes the form of a tiny transformer, in which a current pulse in one direction along the primary circuit magnetizes the core to $+B_{sat}$, leaving a relatively large remnant flux B_r which is held permanently. The orientation of magnetization can be determined by another current pulse, which brings the core to $-B_{sat}$. The voltage appearing on the secondary when this change occurs will be large if the core is at $+B_{sat}$ (logic '1' say) and small if the core is already at $-B_r$ (logic '0' say).

For the best discrimination between $+B_r$ and $-B_r$, a sharp change of state, and long retention of B_r, a square hysteresis loop is required, as illustrated in Fig. 4.18(c).

Switching speed is also important and eddy currents cause a delay. Therefore, ferrites which can be made with a square loop characteristic and which have low electrical conductivity are normally used, although thin metal films, where a single domain extends throughout the entire film thickness, are also used.

Part 3

Phonons and Photons

'Iteration, like friction, is likely to generate heat instead of progress:'

('The Mill on the Floss', by George Eliot)

*'Nature and Nature's laws lay hid in night.
God said,* Let Newton be!, *and all was light.'*

(Alexander Pope)

5. Heat in solids

5.1 Introduction

We did not get far into the book before thermal vibrations, heat in solids, had to be mentioned. The magnitude of the thermal energy decides whether bonding and the formation of solid crystal structures takes place at all. If the thermal energy is too high, the atoms dissociate into a liquid or gas. The crystal structures that do form may undergo a change when the thermal vibration reaches a certain limit.

Thermal vibrations interact with the valence electrons in metals to raise the energy of those electrons near to the Fermi level. In semiconductors, electron–hole pairs are formed when electrons are thermally excited across the band gap, greatly increasing conductivity.

Thermal vibrations limit the mean free path of electrons, as described by Ohm's law and the concept of electrical resistance.

Thermal vibrations fight againt any kind of order in the crystals, such as alignment of magnetic moments. Thus, it is difficult completely to align magnetic moments of paramagnets at temperatures that are not close to absolute zero, and ferromagnetism, the alignment of magnetic moments by exchange forces, breaks down above the Curie temperature.

We will now briefly describe lattice vibrations and introduce the quantum of vibrational energy relevant to elastic waves in a solid. The quantum of heat energy is called a phonon (by analogy with the photon, the quantum of electromagnetic energy or light). The models developed will then be used to explain the variation of specific heat and thermal conductivity with temperature, thermal expansion, and the difference between thermal conduction in metals and non-metals.

5.2 Quantization of lattice vibrations (phonons)

In discussion of the energy distribution of lattice vibrations, we will use many concepts that have already been used in describing electron energy levels in solids. The reason for this is that both forms of energy are wave-like in nature;

additionally, they are quantized and the quantization implies particle-like properties as well.

(a) The vibrating atom

First, consider an atom, say hydrogen, bonded to a second atom. This atom will sit in a potential well, as already illustrated in Figs. 1.8 and 3.5, but reproduced here in Fig. 5.1.

We can think of the interatomic bonding as a strong spring connecting the two atoms. At any real temperature, the atoms will be vibrating and will, therefore, be bouncing around in the potential well. For small amplitudes of vibration (low thermal energies), the walls of the well climbed during oscillation are symmetrical and the atoms have simple harmonic motion – the vibration is said to be harmonic. For high energies, the amplitude becomes large compared with interatomic separation and the well is certainly not symmetrical. There is a very sharp (hard-sphere) repulsion on the side closest to the neighbour due to repulsion of the two nuclei, whilst the potential hill away from the neighbour is much more gentle, rising to a limiting value at large separations where the atom is free. Thermal vibration, where the non-symmetrical part of the potential well is used, has a contribution to the vibrational motion that is not harmonic; this is called anharmonic vibration.

A lot of important calculations can be done assuming harmonic vibration, but many effects, such as thermal expansion, would not occur without the anharmonic contribution.

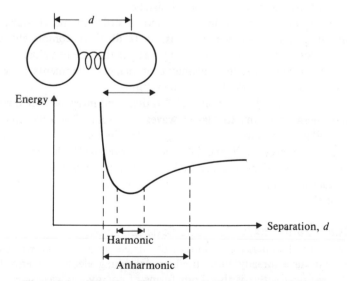

Figure 5.1   Vibration of an atom in the bonding potential well.

Like all bound systems with wave-like properties, the energy of lattice vibrations is quantized and is given by $E_{\text{thermal}} = h\nu(n + \frac{1}{2})$ + anharmonic terms, where h is Planck's constant, ν is the frequency of lattice vibrations (which may be up to 10^{14} Hz), and n is the vibrational quantum number. It is interesting to note that at absolute zero of temperature, where $n = 0$, the atoms still have vibrational energy $h\nu/2$; this is called zero-point energy and arises from the refusal of atomic systems to be completely particle-like. The Heisenberg uncertainty principle states that we cannot know the position and momentum of an atom at the same time. If the atom had zero momentum, this principle would not hold, so there is a certain spread in momentum (and position) which leads to a non-zero energy at absolute zero of temperature.

(b) Vibration of a row of atoms

Let us now consider a row of atoms connected by bonds, shown as springs in Fig. 5.2. This is somewhat of an oversimplification, as atoms are not only bonded to their nearest neighbours, but also to their next-nearest neighbours and even more remote atoms, perhaps as far as 20 lattice constants away. The model of a chain of balls connected by springs is good enough, however, to illustrate some important points. Each atom can have three modes of vibration: two out to the side, the transverse modes, and one backwards and forwards along the line of the chain, the longitudinal mode. The modes we have discussed for two bonded hydrogen atoms are longitudinal modes. As energy is quantized, this string of atoms will have a series of allowed modes of vibration — just as we explained in Chap. 1 for a transverse mode of a violin string. Each mode of vibration will be like an electron stationary state, and we can even plot a probability density for atom position as we plotted the probability density for electron stationary states in Fig. 3.2. This is illustrated in Fig. 5.3 for a mode of quantum number $n = 7$. There are seven most probable atomic positions and they approximate to the probability density of a classical harmonic oscillator,

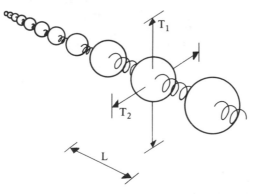

Figure 5.2   Vibrational modes of a string of atoms. T_1 and T_2 are the transverse modes, L is the longitudinal mode.

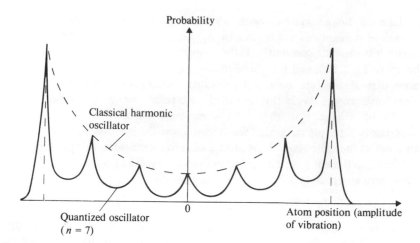

Figure 5.3 The probability of finding the vibrating atom as a function of the amplitude of vibration for a classical harmonic oscillator and a quantized harmonic oscillator for $n = 7$.

which peaks at the extremes of oscillation where the particle is travelling slowly, and where the direction of motion reverses.

We can see that, in a quantized vibration, the atom has only a limited choice of displacements; but when n becomes large, the probability distribution approaches the classical continuum of allowed energies (as in the case for electron energy levels). Each allowed mode of vibration is called a phonon.

(c) Energy, momentum, and group velocity

We know that if we hold a rope taut and give one end a sharp wiggle, energy in the form of a wave will travel along the rope to the other end; if that end is held firmly, the wave may be reflected back. In the same way, energy can be made to travel along our chain of atoms. As they are connected by springs, this can be done equally well using a wiggle — inducing transverse modes — or a jerk — inducing a longitudinal mode. The relationship between energy and momentum (proportional to $1/\lambda$) for thermal vibrations of our chain is plotted in Fig. 5.4. If you remember, we plotted this relation for electrons in Fig. 3.16. The curve has a different shape for lattice vibrations (it is derived from a sine function rather than from $E \propto (1/\lambda)^2$, as is the case for electrons), but there are some similarities, as we shall see.

The spectrum of wavelengths goes from a maximum of twice the chain length (close to zero energy in Fig. 5.4(a)) down to twice the interatomic separation, $\lambda = 2d$. Each mode or phonon between these two extremes is called a normal mode. This is directly analogous to almost free electrons in a crystal lattice. The curve is plotted for both positive momentum and negative momentum, as energy can travel either way along the chain. The region between $\lambda = -2d$ and $\lambda = +2d$ is

133

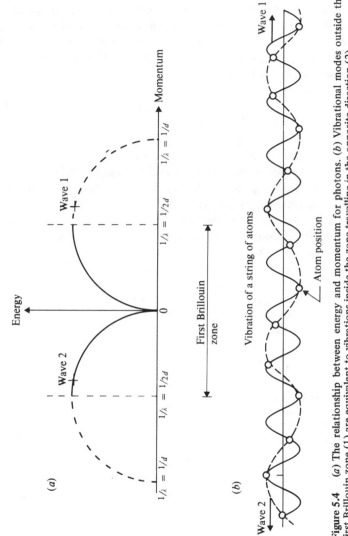

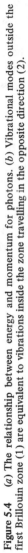

Figure 5.4 (*a*) The relationship between energy and momentum for photons. (*b*) Vibrational modes outside the first Brillouin zone (1) are equivalent to vibrations inside the zone travelling in the opposite direction (2).

the first Brillouin zone. Any vibration of a wavelength that is outside the first zone does not have to be considered, as it is equivalent to a wavelength inside the zone. This is illustrated in Fig. 5.4(b), where a wave of wavelength just less than $2d$ can be seen to be equivalent to a wave of wavelength just greater than $2d$ travelling in the opposite direction.

The analogy with the electron energy—momentum relationship can be carried further. If we look back to the rope-wriggling exercise to transmit energy, energy can be transmitted by thermal vibrations in a similar way in the form of wave packets, i.e., travelling phonons. The conduction of heat by insulators is due to transmission of energy by phonons.

At low frequencies, say up to $\nu \simeq 10^5$ Hz, where the wavelength is of the order of the chain length, the velocity of the wave packet is the velocity of the wave along the chain; this is called the velocity of sound in the solid medium, as it is the speed at which pressure waves propagate. At the highest allowed frequency, where $\lambda = 2d$ and $\nu \simeq 10^{14}$ Hz, the wave packets do not propagate at all; we have standing waves, just as we do for Bragg scattering of electrons at the edge of the energy gap in the electron band structure. The periodicity of the crystal lattice creates a resonant condition.

We can infer, therefore, that the velocity of the wave packets, called the *group velocity*, decreases as the frequency increases, to become zero at $\lambda = 2d$. This phenomenon is known as dispersion and is illustrated in Fig. 5.5. One can visualize the wave packet where the wavelength is close to resonance as a standing wave, created by interference of high-frequency waves; the region where this has an appreciable amplitude moves slowly, whilst the much faster high-frequency waves run through the region but only have an appreciable amplitude in the localized region that is the wave packet.

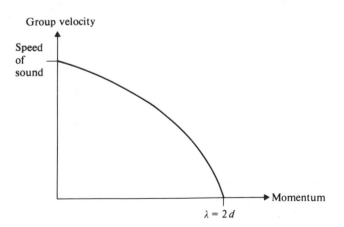

Figure 5.5   The variation of the group velocity of phonon wave packets with momentum.

(d) Real solids

In any real solid, there are a very large number of modes of vibration. These modes range in wavelength from twice the size of the crystal, where they are called sound waves because they propagate at the speed of sound, through the ultrasonic region (10^5 to 10^9 Hz), then to the higher energy almost-standing waves ($\nu \geqslant 10^{11}$ Hz), which contribute most to the thermal energy of the lattice.

The vibrations will be in three dimensions, and, for each direction in the crystal, there will be three sets of modes, corresponding to the three types of oscillation illustrated in Fig. 5.2.

Each normal mode which, remember, is an oscillation of the *whole crystal*, has an energy (neglecting anharmonic terms) of $(n + \frac{1}{2})h\nu$ and is called a *phonon*.

We have indicated that, for a real crystal, the normal modes will be many and complex. There is one further complication that can be mentioned here.

If the crystal has more than one atom per unit cell, then there are more degrees of freedom; several modes which have the same wavelength and type (transverse or longitudinal) can exist by vibrating the different atoms (they may be identical, but with different crystallographic location giving a different force constant to the springs that simulate the atomic bonds) in a number of ways. An example where there are two atoms per unit cell is shown in Fig. 5.6 for a transverse mode of vibration.

In one mode, all the atoms move together as if they were all identical; in the second mode, the two types of atom form two separate chains vibrating anti-phase with each other. This second type of vibration has a different relationship between energy and momentum, forming an upper branch to the energy—momentum curve, as shown in Fig. 5.6. A mode of this second type of vibration is called an optical phonon, whilst a mode of the first type, which is the same as considered earlier for a crystal with one atom per unit cell, is called

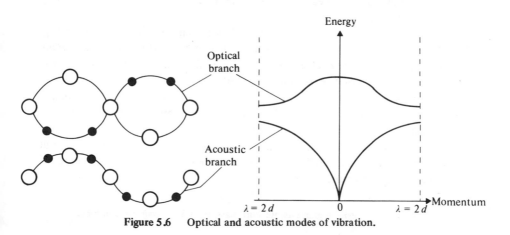

Figure 5.6   Optical and acoustic modes of vibration.

an acoustical phonon. The acoustical phonons are so called because the maximum velocity at which they propagate is the speed of sound. If the two atoms differ in that they carry opposite charge, as in an alkali halide crystal, they can be excited to vibrate in opposition to each other by the electrical field of a light wave; hence the name optical phonon.

If there are more than two atoms per unit cell, there will be an extra set of optical phonons (two transverse and one longitudinal) for each extra atom in the unit cell.

One can see by inspection of Fig. 5.6 that there is an energy gap for phonons, as there is for electrons, when $\lambda = 2d$ and the resonant condition occurs. Any wave with an energy within this gap will be damped and rapidly decay to zero amplitude within the solid.

5.3 The heat capacity of solids

(a) The phonon specific heat

The specific heat is defined as the energy required to raise the temperature of unit mass (usually the atomic weight in grams) by one kelvin. Temperature is related to the degree of vibration of the phonons that are thermally excited. If one could calculate the distribution of phonons as a function of energy by considering every mode in every direction of the crystal structure, then the way that energy is shared by the phonons could be determined analytically and so we could exactly determine the heat energy E_{thermal}. As a result, we could obtain the specific heat at constant volume $C_V = dE_{\text{thermal}}/dT$ at any temperature. This is a very difficult thing to do in practice because of the vast number of phonons that can exist, so some form of approximation has to be made.

The classical approach of Dulong and Petit was that, if we have N atoms free to vibrate at any frequency in one of the three directions in space, then the total thermal energy is given by $E_{\text{thermal}} = 3NkT$, where k is Botzmann's constant. Therefore, the specific heat is independent of temperature; $C_V = 3Nk = 24.9 \text{ J K}^{-1} \text{ mol}^{-1}$.†

The actual variation of the specific heat with temperature is shown in Fig. 5.7. At low temperatures, the specific heat falls to zero. This is the most direct evidence there is of the quantization of lattice vibrations; in other words, it is evidence for the existence of phonons. At low temperatures, only a few low-frequency modes will exist, so a small amount of energy will greatly change the vibrational state of the crystal lattice, thereby causing a large change in temperature. The addition of a quantum of energy makes a large difference when only a few quanta exist; in contrast, in a classical system, an oscillator will accept any amount of additional energy independently of the degree of oscillation already present.

---

†If we use the atomic weight in grams as our unit mass, then N is Avagadro's number.

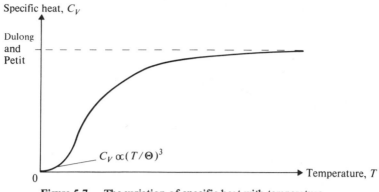

Figure 5.7   The variation of specific heat with temperature.

Einstein demonstrated the existence of phonons by using a very simple model, where he made the gross assumption that the quantized oscillators had only one frequency.

This model was only devised to show that mechanical oscillators were quantized in a way analogous to the quantization of light, established by Planck. The Einstein model predicted the fall in specific heat at low temperatures and fitted the curve quite well, except that the fall predicted at very low temperatures was too steep because, in reality, the low-frequency modes dominate at low temperatures and so there are more phonons available than Einstein's theory would predict.

An approach that yields results more close to the measured values is that due to Debye. He assumed that the solid was a jelly-like structureless block capable of being excited into a series of vibrational modes (in reality, the quantum states or phonons, but not named as such) and that the frequency spectrum of these modes went from zero up to some maximum cut-off frequency, $\nu_{max} = k\Theta/h$. The parameter Θ is called the Debye temperature and is a property of the particular crystal investigated. It is fairly clear now, from inspection of the energy (or frequency)—momentum relationship for phonons given in Fig. 5.4, that this cut-off frequency occurs at $\lambda = 2d$, where we have resonance at a wavelength related to the atomic separation in the crystal and where the phonons have zero group velocity.

Debye used the concept of the density of phonon states $N(\nu)$. This is the number of phonons as a function of frequency, where

$$N(\nu) = N(3\nu)^2/(\nu_{max})^3.$$

Thus, $N(\nu) \propto \nu^2$, as shown by curve A in Fig. 5.8. Again, we have used a concept already introduced for electron energy levels; you will remember that in Fig. 3.3 we plotted the density of states versus energy for electrons (although the density of states is related to energy squared for phonons, not the square root of energy as it is for electrons).

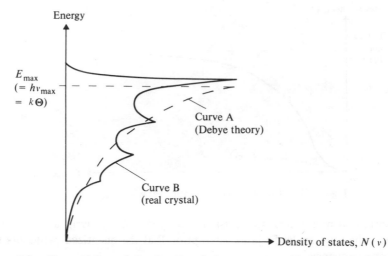

Energy

E_{max}
$(= h\nu_{max}$
$= k\Theta)$

Curve A
(Debye theory)

Curve B
(real crystal)

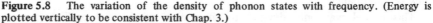

Density of states, $N(\nu)$

Figure 5.8 The variation of the density of phonon states with frequency. (Energy is plotted vertically to be consistent with Chap. 3.)

Debye used statistical mechanics combined with his density of states to predict the variation of specific heat with temperature. The agreement of the Debye model with experiment is remarkable for a comparatively simple model that takes no account of the variation of interatomic separation with direction which occurs in real crystalline solids. In particular, at low temperatures, the model predicts that $C_V \propto (T/\Theta)^3$, which indeed is the case, and also that, at very high temperatures, anharmonic vibration will result in the specific heat being raised slightly above the classical Dulong and Petit value.

The actual density of states in a crystalline solid could be measured once beams of low-energy neutrons become available from nuclear reactors. The neutrons are neutral, and, therefore, when a beam of them are incident on a solid, they are not deflected by the electrons or the electric field of the nucleus. They are scattered by the atomic nuclei and the change in momentum and direction of the scattered neutrons will tell the experimenter the range of atomic positions and thermal momentum in the solid. A density of states can then be calculated. The frequency spectrum for one of our 'model' metals (sodium) measured in this way is shown as curve B in Fig. 5.8.† The density of states is qualitatively similar to the Debye model, but several spikes appear, corresponding to a preference for particular vibrational modes by the crystal structure.

†From Dixon, A. E., A. D. B. Woods, and B. N. Brockhouse, *Proc. Phys. Soc.*, **81**, 973, 1963.

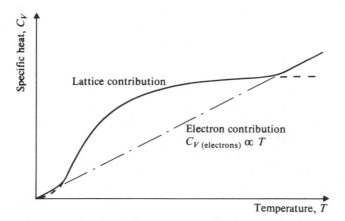

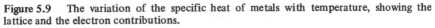

Figure 5.9 The variation of the specific heat of metals with temperature, showing the lattice and the electron contributions.

(b) The electron specific heat

We have already stated that the conduction of heat in insulators is by means of the normal modes of lattice vibration – the phonons. In metals, the conduction of heat is more efficient because this is done by electrons, which can acquire energy from one phonon and transmit this to another phonon.

The thermal conductivity of metals is, therefore, greatly influenced by the number and distribution of free electrons (and we will discuss this further in Sec. 5.4), but they only have a small effect on the specific heat.

We have already touched on the reason for this in Sec. 3.8, where we mentioned the success of the classical free-electron model in determining the relationship between electrical conductivity and thermal conductivity in metals, and the failure of this model to predict the low electron specific heat. This failure is due to the quantization of electron energy and the Pauli exclusion principle, which leads to the Fermi energy distribution of electrons in solids, as described in Chap. 3. Only the most energetic electrons close enough to unoccupied states, i.e., within $\sim kT$ of the Fermi energy, can be excited by thermal vibrations and accept heat energy. It is only these electrons that contribute to the specific heat of the solid. The electron contribution to the specific heat is directly proportional to temperature. (This linear dependence of electron–phonon interaction on temperature has already been described in Sec. 3.12(d)).

If the electron contribution is superimposed on the variation of specific heat with temperature due to lattice vibrations, we have the situation shown schematically in Fig. 5.9.

The electron specific heat is only significant at very low temperatures, where $C_V \propto T$ rather than $\propto T^3$, as in insulators, and at very high temperatures (usually above the melting point of the metal).

5.4　Thermal conductivity in metals

The thermal conductivity of a solid is defined as the power (energy per unit time) that is transmitted through unit area and thickness per kelvin of temperature difference. In metals, the energy is transmitted by free electrons that receive energy from phonon scattering centres nearer the hot side and transfer this energy to phonon scattering centres nearer the cold side. The contribution from the direct transport of phonons is small. We have met this interaction between electrons and phonons earlier, in Sec. 3.12(d), where we saw that electrical resistivity arose from the scattering of electrons by a local variation in interatomic separation d. This moves the band edge so that the electron energy is locally within the forbidden range of energies; in other words, in the band-gap. Thus, the electrons are accelerated by the field; they are then strongly scattered and so reach a limiting velocity, just as friction limits the velocity of a puck accelerated by gravity down an inclined plane. The final velocity of the puck depends on the angle of inclination. The electrical analogue of this behaviour is Ohm's law.

Electrical and thermal conductivity are intimately related in metals as both involve interactions between electrons and phonons. The difference between the two conductivities arises from the fact that electrical charge, the quantity that is transported in an electrical field, is always of the same value. The quantity of heat carried by electrons depends not only on the temperature gradient, but also on the heat energy that can be absorbed by electrons, i.e., the electronic specific heat $C_{V(\text{electrons})}$. We have already stated that $C_{V(\text{electrons})}$ is proportional to temperature; thus, the thermal conductivity K is related to electrical conductivity σ by the Weidemann–Franz law, $K/\sigma = LT$, where L is a constant (the Lorenz constant).

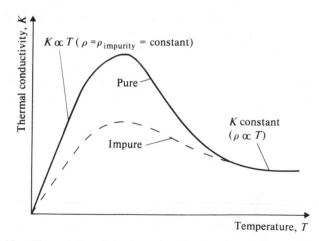

Figure 5.10　The variation of the thermal conductivity of a metal with temperature.

Using this relationship and the already described variation of resistivity ($\equiv 1/\sigma$) with temperature, as shown in Fig. 3.24(c), we can infer that the variation of thermal conductivity with temperature must look like Fig. 5.10.

At low temperatures, resistivity is temperature independent as the impurity distribution is the main scattering mechanism which determines the electron mean free path. The thermal conductivity follows the electron specific heat and is proportional to temperature. At high temperatures, the mean free path is limited by electron–phonon scattering; resistivity is proportional to temperature, but the reduction in electron mean free path is balanced by the rise in electron specific heat, so the thermal conductivity tends to a constant value. Between these two extremes, we have a peak in the thermal conductivity, and the peak height increases as the concentration of impurities that limit the mean free path of electrons at low temperatures decreases.

5.5 Thermal conductivity in non-metals

The relationship between electrical conductivity and the concentration of free electrons in a solid is illustrated in Fig. 5.11.† The free electron concentration can vary by a factor of 10^{24} and so the electrical conductivity will vary by the same degree. This is probably the widest range of any property of matter. The thermal conductivity does not vary by nearly the same amount, only by a factor of 10^3 or 10^4 at the most. The reason for the relatively high value of K for semiconductors and insulators is that the lattice contribution, direct transport of heat energy by phonons, becomes important. The lattice contribution depends on the phonon distribution and not the free-electron density. There is no net flow of phonons during thermal conduction; transfer of energy from the hot end to the cold end occurs by scattering and exchange of energy between phonons.

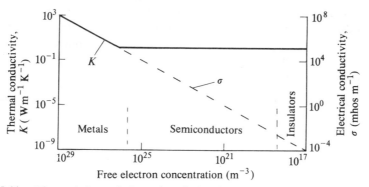

Figure 5.11 The variation of thermal and electrical conductivities with free-electron concentration.

†From Sproull, R. L., *Sci. Am.*, 207, (No. 6) p. 101, 1962.

The value of thermal conductivity in non-metals is dependent on three factors.

1. The number of phonons, i.e., the specific heat.
2. The group velocity of the phonons, i.e., the velocity of sound in the crystal, which depends on atomic mass and bond strength (the spring constant).
3. The mean free path of phonons between scattering events.

The main ways in which phonons can be scattered are as follows.

1. *Direct phonon–phonon interaction* If the modes are harmonic, they cannot interact; we can only get phonon–phonon scattering if the vibrations are anharmonic. This is a well-known principle of electronics: one waveform can only modulate another when we have some form of non-linearity. As phonons are harmonic at low temperatures, phonon–phonon scattering will only occur at high temperatures. Two phonons can then interact to create a third. If the third phonon has a wavelength outside the first Brillouin zone, $\lambda < 2d$, we have the situation shown in Fig. 5.4, where this is equivalent to a phonon of $\lambda > 2d$ travelling in the opposite direction. We have reversal of momentum in a way analogous to the scattering of electrons during electrical conduction. This type of phonon–phonon scattering is called an Umklapp process or U process for short.
2. *Scattering due to inhomogeneities in the crystal structure* These may be point defects, dislocations in the crystal, or variation in atomic mass due to the presence of impurities or a number of isotopes in the crystal structure. These inhomogeneities cause a disturbance in the movement of the phonon wave packet due to the creation of a local situation where reflection (reversal of momentum due to the wavelength being outside the first Brillouin zone) may occur as a result of variation in the interatomic separation or the vibrating mass constant. This scattering becomes small when the phonon wavelength becomes large compared with the disturbance; hence, scattering due to inhomogeneities becomes insignificant at very low temperatures.
3. *Boundary scattering* A surface, either the true surface or a grain boundary within a polycrystalline solid, ultimately limits the phonon mean free path for long wavelengths. Thus, boundary scattering reduces the thermal conductivity at very low temperatures.

The various processes that affect the variation in thermal conductivity with temperature for semiconductors and insulators are summarized in Fig. 5.12. This is similar to the variation of K with T for metals, but it arises from different mechanisms.

The reduction of mean free path due to U processes dominates the thermal conductivity at high temperatures.

At low temperatures, the dominant mechanism is the rapid fall in the number of phonons because $C_V \propto T^3$ and thermal conductivity drops rapidly with the fall in specific heat. Boundary and inhomogeneity scattering are significant at

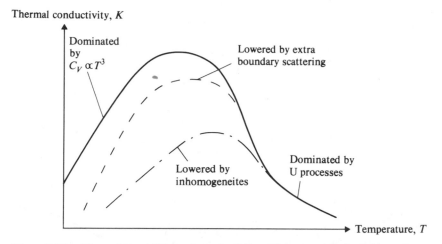

Thermal conductivity, K

Dominated by $C_V \propto T^3$

Lowered by extra boundary scattering

Lowered by inhomogeneites

Dominated by U processes

Temperature, T

Figure 5.12 The variation of thermal conductivity with temperature for non-metals.

temperatures between the two extremes, lowering the value of thermal conductivity from its peak value. Boundary scattering is more important on the low-temperature side of the peak, whilst inhomogeneities have a greater effect on the high-temperature side.

5.6 Practical aspects of heat in solids

(a) Thermal expansion

Thermal expansion is another phonon property that (like the U process) arises from the anharmonic component of lattice vibration. If the atom is vibrating in a deep potential well, the vibration is harmonic and the mean position of the atom is independent of the amplitude of vibration; therefore, the lattice spacing remains the same – there is no thermal expansion. This is illustrated in Fig. 5.13(*a*). If the atom is vibrating in a shallow potential well, as shown in Fig. 5.13(*b*), then the vibration is anharmonic and the mean interatomic separation will increase as the amplitude of vibration increases – the crystal lattice expands. The degree of thermal expansion depends on the shape and depth of the potential well, which, in turn, depends on the nature and strength of the interatomic bonds.

The bond strength (the depth of the potential well) can be gauged from the melting point. Metals with low melting points like lead and tin (600 °C and 505 °C, respectively) will have a large anharmonic component at room temperature and, therefore, a high thermal expansion coefficient – up to seven times the expansion rate of a high melting point metal like tungsten (T_m = 3680 °C).

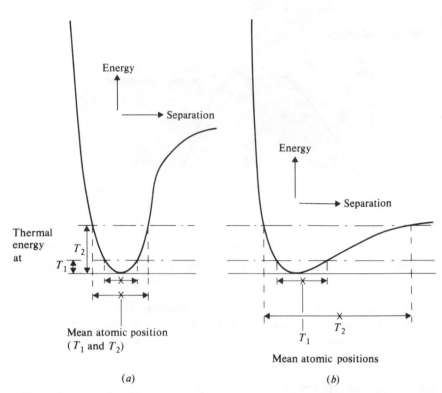

Figure 5.13 Comparison of mean atomic positions. (*a*) Deep potential well, harmonic oscillation, no expansion; (*b*) shallow potential well, anharmonic oscillation, appreciable expansion.

The lowest expansion coefficient metals are alloys such as Invar, which has an expansion rate one third of that of tungsten.

The lowest expansion coefficients occur for non-metals like fused quartz, a glass formed from SiO_2. This has a thermal expansion coefficient that is two orders of magnitude smaller than tungsten. In fact, the thermal expansion coefficient α is related to the specific heat C_V and, therefore, has the same temperature dependence. For many solids, these two parameters and the compressability of the solid, X (which is determined by the shape on the repulsive side of the potential well), are related by the expression

$$\gamma = \alpha V_0 / C_V X,$$

where γ is a constant which is independent of temperature ($\gamma \simeq 2$ for many solids), and V_0 is the volume of the solid.

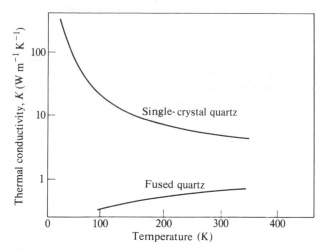

Figure 5.14   Variation of thermal conductivity with temperature for fused quartz and for single-crystal quartz.

(b) Thermal conductivity

The materials with the lowest thermal conductivity and that, therefore, make the best insulators are the glasses. In Fig. 5.14,† the variation of thermal conductivity K with temperature for single-crystal quartz and for fused quartz (a glass formed from a random arrangement of SiO_4^{4-} tetrahedra) are compared. For the single crystal, the value of K is relatively high but drops as temperature increases due to reduction in phonon mean free path by U processes. The glass, on the other hand, has a much lower value of K and show a slight rise of K with increasing temperature following the change in specific heat with temperature. The phonon mean free path in glasses is very small and is independent of temperature. It is determined by the crystallite size, which, as a glass has no regular crystal structure, is the size of the SiO_4^{4-} tetrahedron. This very strong phonon scattering yields a value of thermal conductivity that is two orders of magnitude less than single-crystal quartz and up to six orders of magnitude less than some metals.‡

It is sometimes desirable to lower the thermal conductivity while retaining reasonable electrical conductivity. One example where this is necessary is the thermopile, which generates electricity by having dissimilar metals joined in two

†Modified from Kittel, C., *Introduction to Solid State Physics*, 4th ed., John Wiley, New York, 1971, p. 231, Fig. 27.
‡The range of K for solids given earlier was four orders of magnitude, but this applied to single crystals.

places. One of the junctions is heated and the other cooled and a thermoelectric voltage is created by the difference in the junction potential barrier at the two temperatures determined by the Fermi levels of the two metals.† Heat is wasted by conduction between the hot and cold ends. Thermal conductivity is reduced by using metal *compounds* where heat conduction is by phonons, doped with a heavy metal such as tellurium. This reduces thermal conductivity by impurity scattering and by reducing the velocity of sound in the solid. Sufficient numbers of free electrons for electrical conduction are supplied by the metal atoms.

† In the thermocouple this voltage is used to measure temperature.

6. Photons in solids

6.1 Introduction

Unlike the case for electrons, we are familiar with the wave-like properties of electromagnetic radiation from the long-wavelength radio waves through the visible spectrum to X-rays and, ultimately, to the most energetic of all, the γ-rays, which are emitted from excited states within the nucleus. Like all other forms of energy though, electromagnetic waves have particle-like properties, i.e., they are quantized and the quanta are called photons. These carry no charge or mass, but have energy and travel at the velocity of light in free space.

That light was corpuscular in nature was first postulated by Newton, but he used classical mechanics and encountered difficulties in explaining with sufficient elegance and simplicity properties like the refraction of light in solid media. Several hundred years later, Planck postulated the photon of energy $h\nu$ (where h is Planck's own constant) in order to explain the spectrum of radiation emitted from hot solids at thermal equilibrium with their surroundings (black bodies). Evidence for the quantization of light comes also from the character of the photoelectric effect, where electrons are emitted from solid surfaces when light is incident upon them. The electrons have energies dependent only on the frequency of the radiation (the energy of the photon) and not on the intensity (the number of photons). One would expect an intensity dependence from classical electromagnetic theory.

This chapter is divided into three sections. In the first section, we will describe the interaction between photons and electrostatic dipoles in insulators: the polarization of dielectrics. In the second section, we will describe the absorption of photons by solids. The third section deals with emission of photons by solids, a subject already touched on several times earlier in this book. For example, in Sec. 3.12(b) we described the emission of soft X-rays for transitions from a broad band to a well-defined energy level. In the following section (Sec. 3.12(c)), the emission of light from semiconductors was seen to occur only when transitions across the band-gap involved no change in momentum of the electrons. When dealing with magnetic moments (see Fig.

4.3), it was observed that splitting of electron energy levels in a magnetic field leads to splitting of spectral lines arising from transitions between these split levels and a lower energy level. These phenomena will be fitted into the general pattern of the emission of photons from solids and practical applicatons such as phosphorescence and lasers will be described.

6.2 Dielectrics

Photons are strongly absorbed by metals and decay to zero amplitude within a short distance of the surface. The thickness of the skin that is affected is of the order of the wavelength of the photon. The reason for this very strong attenuation is that there is a quasi-continuous band of allowed states in the region of the Fermi level. Electrons can be excited to vacant states by interaction with the incident photons. Most photon energy is reflected and this is why polished metal surfaces make good mirrors.

Photons of energy less than the band gap (with certain exceptions detailed in Sec. 6.3) can travel through insulators without attenuation: the crystals are transparent. Therefore, interactions other than absorption by excitation of electrons become important.

Classical theory tells us that the refractive index of any material is given by

$$n = \frac{\text{Velocity of light in vacuum}}{\text{Velocity of light in the medium}}.$$

The velocity of light in a medium is $(\mu_0 \mu_r \epsilon_0 \epsilon_r)^{-1/2}$, where ϵ_0 is the permittivity of free space and ϵ_r is the relative permittivity. If we have a non-magnetic material, then $\mu_r = 1$ and the refractive index $= \sqrt{\epsilon_r}$. The relative permittivity, and, therefore, the refractive index, is increased if dipoles exist or can be induced in the solid. In other words, the dipoles reduce the group velocity of the wavepackets of electromagnetic energy, the photons.

There are many analogies between dielectric behaviour and magnetic behaviour and we will use these in describing dielectrics. Permittivity is analogous to permeability and we can say that an externally applied electrical field E induces an electrical effect in the solid, the electric displacement D being given by $D = \epsilon_0 \epsilon_r E$ (by analogy with $B = \mu_0 \mu_r H$). The field causes polarization P of the solid. This polarization is defined as the dipole moment per unit volume and is, therefore, analogous to the magnetization M. A practical measure of how easily a solid is polarized is the change in capacity of a capacitor when the dielectric is inserted. This is related to the amount of polarization induced in the solid by a certain value of electric field. There are three ways a solid may be polarized and these are illustrated in Fig. 6.1.

We have already met electronic polarization (Fig. 6.1(a)) when discussing van der Waals bonds. The polarization involves displacement of the electron cloud around the nucleus by the electric field. Ionic polarization (Fig. 6.1(b)) involves the movement of ions of opposite charge with respect to each other. The ionic

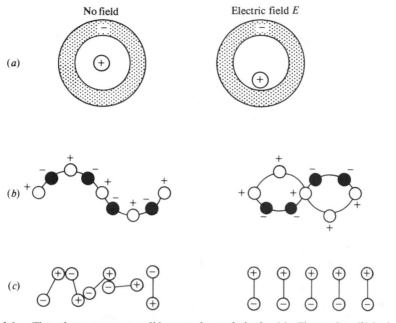

Figure 6.1 The three ways a solid may be polarized. (*a*) Electronic; (*b*) ionic; (*c*) molecular.

polarization illustrated is the transverse optical phonon mode (discussed in Sec. 5.2 and shown in Fig. 5.6) applied to the case of an ionic crystal. The molecular dipoles (Fig. 6.1(*c*)) are permanent dipoles that exist in the solid. For example, in the water molecule the bonding electrons spend most of their time in the vicinity of the oxygen atom, which, therefore, is negatively charged. This leaves a net positive charge on the two hydrogen atoms and the whole molecule is a dipole. An external electrostatic field will tend to align the permanent molecular dipoles in a way analogous to the alignment of magnetic moments in a magnetic field.

The effect of an incident photon is to subject the electrical dipoles in a solid to an alternating electric field at a frequency given by the photon energy $E = h\nu = h\omega/2\pi$.† Molecular dipoles take time to respond to an electric field and this response will have a characteristic time constant τ. The value of τ depends on the nature of the dipoles, but, typically, $\tau \simeq 10^{-11}$ s. As we increase the photon frequency, the dipoles start to get left behind by the electric field. The

---

† Often, frequency is quoted in rotational terms, i.e., in radians per second, this is given the symbol ω and, as there are 2π radians for each cycle, $\omega/2\pi = \nu$, where the units of ν are hertz.

150

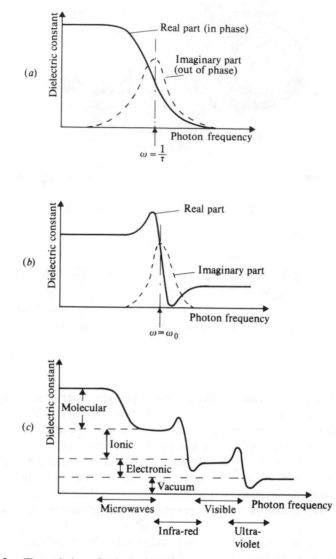

Figure 6.2 The variation of polarization with frequency. (*a*) Dipole relaxation; (*b*) resonance absorption; (*c*) characteristics of a material exhibiting all three types of polarization.

displacement D has a component that is out of phase with the electric field E. The magnitude of this out-of-phase component is a measure of the power dissipated in the solid and is at a maximum when the photon frequency $\omega = 1/\tau$. This phenomenon is known as dipole relaxation. It is just this phenomenon that is used to cook food by bombarding it with photons of wavelength $\lambda \simeq 10$ mm in a microwave oven. The photon energy is absorbed by the movement of the dipoles and converted into lattice vibrations — phonons.

The electrical permittivity is sometimes called the dielectric constant ‡ and is related to the amount of polarization that can be induced by an electric field. The variation of dielectric constant with photon frequency for molecular dipoles is shown in Fig. 6.2(a). At low frequencies, the dipoles can easily follow the field and the dielectric constant is equal to the static value $\epsilon_0 \epsilon_r$. As the frequency approaches $\omega = 1/\tau$, the in-phase (real) component reduces in magnitude whilst the out-of-phase (imaginary) component increases to a maximum at $\omega = 1/\tau$. At frequencies where $\omega \gg 1/\tau$, the dipoles have insufficient time to respond to the field, there is no overall polarization, i.e., zero contribution to the dielectric constant from permanent dipoles, and, therefore, there is no power loss.

The mechanism that determines the change of dielectric constant with frequency for dipoles that are *induced* by the electric field, i.e., electronic and ionic dipoles, is quite different to that for permanent dipoles, although the result looks rather similar. Induced dipoles are bound vibrating systems — such as an electron bound to a nucleus by Coulombic attraction. As explained in Chap. 1, this is a resonant system. There is a natural resonant frequency ω_0. (There will also be harmonics of this natural frequency, but for simplicity we will just consider the fundamental.)

A photon will subject this natural resonator to an alternating electric field. The variation of dielectric constant with photon frequency is shown in Fig. 6.2(b). At low frequencies, the polarization follows the field very easily and the dielectric constant is the same as for a constant field. As the frequency approaches ω_0, the induced dipole starts to resonate. Vibrations, forced by the external driving field, tend to build up and the dielectric constant tends to increase, but the displacement starts to get out of phase with the electric field. As in the case of dipole relaxation, it is the out-of-phase component that causes energy loss by scattering the photons, with conversion of the photon energy into phonons (heat). The energy loss is a maximum when the photon frequency equals ω_0. The real part of the dielectric constant is greatly reduced because of the large amount of photon energy that contributes to the imaginary part and is converted into heat. For ionic dipoles, ω_0 is in the infra-red wavelength region and it is this mechanism of resonant energy loss that we know as infra-red heating: the warmth we feel from a fire or from the sun.

At frequencies much greater than ω_0, there is insufficient time for a dipole to be induced at all, and the contribution to the dielectric constant from these dipoles disappears, as does the photon energy loss.

‡Sometimes, this term is used for the constant of proportionality between polarization P and field E; in this case, the dielectric constant is analogous to magnetic susceptibility χ. We will keep to the more exact definition, although the two definitions are similar for $\epsilon_r \gg 1$ (just as μ_r and χ are similar for $\mu_r \gg 1$).

A solid may contain all three forms of polarization – molecular, ionic, and electronic. A schematic diagram of the variation of dielectric constant with photon wavelength for such a solid is shown in Fig. 6.2(c). The molecular contribution is lost at microwave frequencies, the ionic in the infra-red, and the electronic at ultra-violet frequencies. The electronic energy loss at the resonant frequency may be energetic enough to break chemical bonds. Anyone who has had sunburn has experienced this form of photon energy loss. The same phenomenon is put to good use in making electronic integrated circuits. In such circuits, silicon, covered by an organic liquid called photo-resist, is irradiated with ultra-violet light through a mask. The exposed regions undergo a chemical change which alters their chemical reactivity. Then, either the exposed regions or the unexposed regions are dissolved away, leaving the desired pattern on the silicon surface. After other processes, this pattern is converted into doped regions in the surface of the semiconductor. This process is called photo-lithography, but conventional photography depends on photon-induced chemical changes also.

Electronic circuits normally operate at frequencies below the microwave region, where all three mechanisms contribute to the dielectric constant and, therefore, to the value of a capacitor. The last thing that we want is the conversion of photon energy into heat as this would, at best, lead to a loss of energy from the circuit. Therefore, we need to use a dielectric that has no dipole relaxation at the frequencies used. We may need a high value of capacitance, i.e., high polarization for a given field. Just as the induction of polarization is analogous to the magnetization of a paramagnet, we also have the analogue to ferromagnetism, which is called ferroelectricity. This has been briefly mentioned in Sec. 2.14(a), where the structure of barium titanate was described (Fig. 2.15). A permanent polarization occurs through displacement of the sublattice of Ba^{2+} and Ti^{4+} with respect to the O^{2-} sublattice. The polarization creates a field which distorts the lattice still further; this, in turn, creates more polarization. Similarity with magnetic behaviour includes the fact that ferroelectric domains can form and ferroelectric behaviour disappears at temperatures above the appropriate Curie temperatures (393 K for $BaTiO_3$). Also, antiferroelectric materials are known to exist.

The ferroelectric effect results in a very high dielectric constant, which is the reason for use of these materials to make high-value capacitors. There are other interesting properties of these materials. For example, the ferroelectric with the highest polarization and Curie temperature (1470 K) is lithium niobate $LiNbO_3$. This is used in devices where the piezoelectric properties of these materials are exploited. Piezoelectricity is the conversion of electrical signals into mechanical vibrations or vice versa. It arises from the fact that any distortion of the crystal structure means a large change in polarization, creating a large change in the charge induced by the electric field on electrodes placed on the surface of the dielectric, conversely an electric field will cause a change in shape by distortion

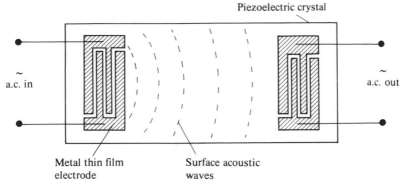

Figure 6.3   Schematic representation of a surface acoustic wave delay line.

of the crystal. One use of piezoelectricity is for ultrasonic drills, but a more interesting application is the use of the effect to make surface acoustic wave (SAW) devices. Electrical signals on an electrode (shown in Fig. 6.3) placed on a thin slice of single-crystal lithium niobate can move the lattice by altering the polarization in phase with the signal. The crystal is excited into vibration (a low-frequency phonon). This vibration propagates along the surface at the speed of sound and can be converted back into an electrical signal by a second electrode. This device has many sophisticated uses. Two fairly simple examples are: (i) use as a delay line, as the propagation speed of lattice vibrations is very low compared with that of electrical signals; (ii) use as an electronic filter, where only frequencies within the allowed modes of vibration (which can be controlled by the geometry of the device) will propagate and be picked up at the output of the device.

Some piezoelectric solids do not have a polarization at zero field – they are not ferroelectric – but they acquire a polarization when stressed. For this to happen, the lattice must have a lack of symmetry in the direction of stress. One example of this class of dielectrics is quartz, crystals of which are used as very stable oscillators. In the oscillators used in watches quartz crystals are excited into a bulk vibration by using electrodes on opposite sides of a single-crystal slice. The low thermal expansion of quartz reduces frequency shifts caused by changes of temperature.

6.3   Absorption of photons by crystals

Insulating crystals are usually clear because the wavelength range of the visible spectrum (7400 to 3600 Å) is equivalent to photon energies between 1.7 eV and 3.5 eV, whilst most insulators have a band gap of greater than 4 eV.

The phenomena that lead to absorption of photons and, therefore, the creation of colour in crystals of insulators are shown schematically in Fig. 6.4.

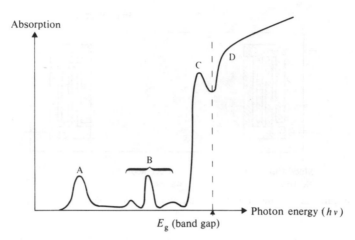

Figure 6.4 The absorption of photons by insulators as a function of energy. A, photon–phonon interaction; B, colour centres; C, exitons; D, excitation across the band gap.

Absorption peak A arises from the resonant interaction between photons and the optical phonon modes of ionic crystals, as discussed earlier when we were concerned with polarization of dielectrics. Energy is absorbed and transmission drastically reduced at the point where the resonance occurs, which is usually in the infra-red. This effect can be used to polarize light reflected from the surface of an alkali halide crystal if the wavelength is such that only one optical phonon mode is excited, thus absorbing strongly only one component of the incident light. The wavelengths at which these resonances occur are called the Reststrahl wavelengths. The effect can also be used for production of lenses and prisms from alkali halide crystals with a high refractive index in the infra-red.

The family of peaks at B represents absorption due to electron transitions between localized energy levels of impurity atoms or defects. For example, the excited states of transition-metal ions are in the visible range of wavelengths. A crystal of alumina (Al_2O_3) is normally transparent but with a trace of chromium as an impurity it becomes the red ruby, whilst, if we have a trace of titanium instead, we have the blue sapphire.

An atom missing from the crystal can also create localized excited states for electrons. For example, if a negative ion is missing from an alkali halide crystal, this leaves a net positive charge in the vacant lattice site which can trap an electron. This system, with an electron loosely bound to the positively charged vacancy, acts rather like a hydrogen atom with excited states for the electron to occupy on interaction with a photon. This particular type of colour centre is called an F centre. An F′ centre is one with two loosely-bound electrons. The V centre is a positive hole trapped at a positive ion vacancy. The vacancies can be created by bombardment with energetic particles or by heating the crystal in the vapour of one or the other of the constituents; new crystal with then grow with a deficiency of one component, which diffuses as vacancies into the solid.

The variation of the probability of photon scattering with photon wavelength for small particles provides another method of achieving colour in insulators. Ruby-coloured glass is made by producing a colloidal suspension of gold in the glass, which, because the size of the particle is of the order of the wavelength of light, scatters the short wavelength (blue) light much more strongly than the long wavelength (red) light. It is the transmitted light that gives colour to the glass.†

The peak at C occurs at photon energies just less than the band-gap energy. The absorption comes from the creation of electron—hole pairs, where the electron does not have enough energy to reach the comparative freedom of the conduction band. The electron and hole remain loosely bound to each other and the pair is called an exciton. The band diagram for this is illustrated in Fig. 6.5(a). The exciton can travel through the crystal, although no net charge is carried, and has a series of hydrogen-like energy levels just below the conduction band edge.

The absorption at D in Fig. 6.4 represents excitation of electrons across the full band gap to create relatively free electron—hole pairs, as shown in Fig. 6.5(b). The semiconductors with small band gaps, for example, silicon, with $E_g = 1.14$ eV, look like metals because all visible wavelengths excite direct

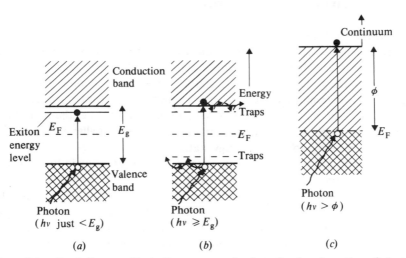

Figure 6.5 Band diagrams illustrating some mechanisms for the absorption of photons. (a) Creation of exitons; (b) photoconductivity; (c) photoelectric effect.

†Whereas the *scattered* light from the sun makes the sky blue.

transitions across the band gap. Cadmium sulphide is a wide-gap semiconductor with E_g = 2.42 eV; only blue light is absorbed, so the crystals are yellow/orange in colour (created by the reflected or transmitted light). The creation of free electrons and holes by interaction with photons is called photoconductivity. This effect is used in television cameras, image intensifiers, infra-red detectors, light meters (cadmium sulphide is often used), and, indirectly, in photography. Localized energy levels due to defects and impurities are very important in determining the performance of photoelectric devices. These are called *traps* and are of two types.

1. *Recombination centres*, where electrons and holes can recombine by exchange of momentum with phonons (especially necessary for indirect-gap semiconductors like silicon and germanium). These affect the lifetime of charge carriers in the device.
2. *Temporary trapping centres*, which reduce the mobility of one or both the charge carriers, and, therefore, determine the conduction type, sensitivity, and response time of the device.

Also illustrated (Fig. 6.5(c)) is the photoelectric effect for metals. This effect is similar to excitation of electrons across the band gap of semiconductors. However, in the case of metals, the excitation is between the Fermi level and the continuum, the photon energy has to be greater than the work function rather than the band gap, and the electron is emitted from the surface of the metal.

6.4   Emission of photons by solids

(a) Spontaneous emission

Photons are emitted from atoms when electrons that have been promoted to an excited state decay directly to a lower state. The photon energy is just the difference in energy between the two states. The spread in photon energy, the 'line-width', is determined by the amount of time that the electron spends in the excited state. The longer an electron stays in an excited state, the more 'well defined' the energy of the electron becomes† and the smaller the spread in energy when it decays to a lower energy state.

Some transitions of excited electrons are shown schematically in Fig. 6.6. The photons are of microwave frequencies when the transitions are between excited states in levels split by an external magnetic field. Light in the visible region, the spectral lines, arises from transitions between excited states and vacancies in the higher energy levels. This will show the magnetic splitting, if it exists, as illustrated in Fig. 4.3. Thermal energy sufficient to vaporize the material of interest (say a temperature of 1000 K) is usually necessary to excite these levels.

---

† 'Switching on' a wave involves frequencies other than the fundamental, but the significance of this gets less as time goes on. This is another consequence of the Heisenberg uncertainty principle.

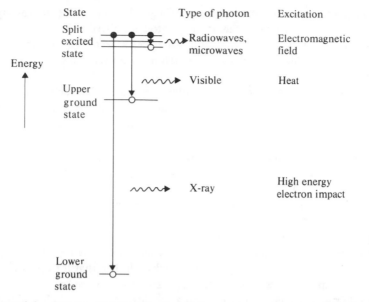

Figure 6.6   Electron transitions leading to the spontaneous emission of photons.

If we bombard a solid with electrons of high energy (say >100 keV), then electrons can be excited from the lowest energy levels. When electrons then decay to fill these vacancies, high-energy photons of X-ray wavelengths are emitted. If the electrons decay from well-defined inner levels, we obtain a series of lines characteristic of the atom. Transitions to the $n = 1$ state are called the K series, transitions to $n = 2$ states are called the L series and so on. It is these X-rays that are used for analysis of crystal structures by Bragg diffraction. Transitions from the valence band to an upper energy level give rise to a wide spectrum of energy, as seen in the soft (low-energy) X-ray spectrum described in Sec. 3.12(*b*).

Spectral lines are emitted by highly excited gas atoms, but visible light of well-defined frequency is also emitted by atoms in solid semiconductors and insulators. This occurs because of the decay of an excited electron from the bottom of the conduction band to occupy a vacant state in the valence band. The energy of the photon emitted by this electron–hole recombination is the band-gap energy E_g. The process is the reverse of the photon excitation across the band gap shown in Fig. 6.5(*b*). In order that a photon is emitted (rather than the excitation of a phonon), it is necessary that the electron does not have to change momentum as well as energy, i.e., we have a direct gap, as described in Sec. 3.12(*c*). This is the case for GaAs and some other compound semi-conductors. One such compound, with the band gap tailored to emit red light, is used to make light-emitting diode (LED) displays in watches and calculators.

The phenomenon of the emission of light by solids due to this type of transition is called luminescence. If the photon is emitted only a short time after excitation (short on the atomic time-scale means less than 10^{-8} s), then the phenomenon is called fluorescence. Emission for longer times after excitation is termed phosphorescence.

Excitation can be by photons (photoluminescence), electron beams (cathodoluminescence) — as used for television and oscilloscope screens — chemical reactions, electromagnetic fields, or d.c. electric fields (as in the gallium arsenide light-emitting diodes or LED).

The compound semiconductor zinc sulphide is a material that is used as a phosphor. You will remember from Sec. 2.12 that the crystal structure is analogous to that of silicon. The band gap is fairly wide, so photons in the visible spectrum are emitted. A few parts per million of copper, silver, or gold are introduced to replace zinc in the lattice. These are called activators as they create trapping levels in the band gap close to the valence band (similar to acceptor levels of dopants in silicon) which facilitate radiative decay (Fig. 6.7). The emitted light has a slightly longer wavelength than that for decay across the full band gap. The decay from the activator trap to the valence band is by interaction with phonons; it is non-radiative. Materials that produce donor-type trapping levels close to the conduction band are called coactivators, they include gallium and indium on zinc sites and chlorine on sulphur sites. These traps act like the traps described when we were discussing photoconductivity. They increase the lifetime of the electron in the conduction band by sequential trapping and release of the electrons by non-radiative energy exchange with the phonons. This increases the persistence of the phosphor.

Luminescence also occurs by electron decay from localized defect and impurity excited states, the reverse of absorption by these states discussed in the previous section. One interesting example of this behaviour is that of thallium in KCl. The excitation occurs at energy $h\nu_1$, but this alters the energy balance in the crystal lattice and the interatomic spacing is changed in such a way that the band gap is reduced. The photon that is emitted on decay of the electron to the

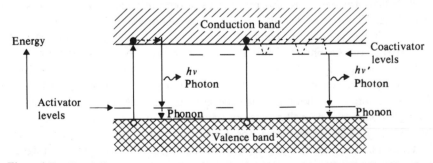

Figure 6.7   Band diagram and localized impurity levels in a semiconducting phosphor.

ground state has a lower energy $h\nu_2 < h\nu_1$. Thus, light of one wavelength is absorbed and the crystal luminesces at a longer wavelength.

The type of decay of an electron from an excited state that we have discussed so far occurs spontaneously at a time determined by the statistical probabilities of the event taking place. This is termed spontaneous emission, and any photon that is emitted has no phase relation to the exciting radiation or to other photons that are emitted. The light is *incoherent*.

We will now discuss the case where an incident photon can initiate the decay of an excited electron. When this occurs, the emitted photon has the same frequency, is in phase with, and has the same propagation direction as the incident photon. In a succession of similar events all the emitted photons can be in phase and the light is said to be *coherent*. This is termed *stimulated* emission.

(b) Stimulated emission

The phenomenon of stimulated emission is illustrated in Fig. 6.8(*a*). An electron in an excited state (atom 1) spontaneously decays to the ground state, emitting a photon of energy $h\nu$ equivalent to the separation of the two energy levels. This spontaneously emitted photon is just the right energy to cause (or stimulate) a second electron in atom 2 to decay to the ground state, with the emission of a second photon that has the same frequency and the same phase as the stimulating photon. The total photon energy has been doubled by this process.

Stimulated emission can be used, therefore, for amplification of photon energy, as illustrated in Fig. 6.8(*b*).

Here, we have assumed that all the electrons are in excited states. The means of achieving something approaching this unusual situation, termed a *population inversion*, will be discussed later. Assuming, therefore, that we have a population inversion, one can see that as the photon wave front passes each electron it stimulates a decay to the ground state with the emission of another photon which is in phase with and of the same frequency as the stimulating protons. We have *l*ight *a*mplification by *s*timulated *e*mission of *r*adiation – in other words a *laser*.

The amplification is increased many times by placing the material in which these transitions occur into a resonant cavity, as illustrated in Fig. 6.8(*c*). In the simplest form, mirrors are placed at each end of a cylindrical cavity containing the laser material, so that the light is reflected backwards and forwards, being amplified on each pass along the axis. Photons that are not directed along the axis or do not have the correct frequency are not amplified. Not only are all the photons in phase and of the same frequency, but they become channelled so that the direction of propagation is very closely parallel to the optical axis. The light is extracted from the laser by designing one of the mirrors to be slightly less than 100 per cent reflective. It is this small loss, together with losses from the walls, from the absorption of photons by excitation of electrons not in excited

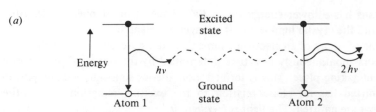

(a)

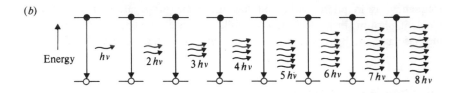

(b)

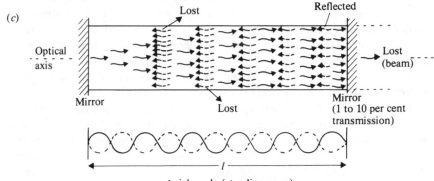

(c)

Axial mode (standing wave)

Figure 6.8 Amplification of light by stimulated emission. (a) Stimulated emission; (b) amplification by successive stimulated emissions; (c) the laser resonant cavity (showing one axial mode).

states, and from spontaneous emission that places a limit on the amplification process.

The laser forms a resonant cavity and, therefore, the photons will form standing waves. The simplest of the modes that can be excited are called *axial modes*. These are analogous to the violin-string modes discussed in Sec. 1.5, where the state (or mode) number $n = 2l/\lambda$. In the case we are considering here, l is the length of the laser, say 0.5 m, and λ is, say, 0.5 μm. Therefore, n is about two million. As n is so large, any slight variation in length of the laser will enable several modes to be stimulated. Transverse modes can also occur to complicate the picture. However, lasers can be made where just a single axial mode is excited by accurate control of mirror separation to compensate for such things as temperature variations.

The light emitted from a single-mode laser is equivalent to a single very long wave packet. This is a consequence of the phase agreement of all the photons and the standing wave pattern in the laser cavity. The light is said to be *coherent*. The *coherence length* is the distance one can go from the laser and still measure a phase relationship, roughly, therefore, the length of the wave packet. Coherence lengths of tens of metres† can be achieved and one can, therefore, predict from the Heisenberg uncertainty principle‡ that the uncertainty in momentum (and, therefore, wavelength) is very small. The laser is a source of very pure, almost single-frequency light, i.e., it has a very small spectral line-width. This can be several orders of magnitude smaller than the line-width for the more normal excited electron energy loss, spontaneous emission, where the minimum line-width is determined by the time spent in the excited state (usually less than 10^{-8} s) and line-width is increased by the relative movement of atoms (doppler broadening) and by the distortion of energy levels due to neighbouring atoms (pressure broadening).

A single-mode laser is not only coherent along the axis of the beam (termed temporal coherence), but also at right angles to the axis (spatial coherence), although this degree of coherence is not necessary for many laser applications.

The high degree of coherence achievable with laser light and the close agreement between the direction of propagation and the optical axis results in a beam with a divergence of the order of milliradians. This cannot be achieved with conventional light sources as these emit light in all directions. Even if light from a non-coherent source is focused into a beam, there is a large divergence due to diffraction at the edges of beam-limiting apertures. Laser light has a maximum energy at the axis and is not constrained by the boundaries of the system, as the energy falls to zero at the edge of the beam. The only divergence that occurs is due to diffraction caused by phase variations within the beam.

We will now turn our attention to the means by which a population inversion can be achieved. Consider two energy levels, as shown in Fig. 6.9(a), where, at zero temperature, level 1 would be occupied and level 2 would be vacant. At non-zero temperatures, some electrons would be excited to level 2 by thermal vibrations, and the number occupying this level (N_2) would be governed by the same rule as quoted when considering the magnetization of solids, in Sec. 4.9 i.e.,

$$N_2 = N_1 \exp(-\Delta E/kT),$$

---

†Hundreds of kilometres can be achieved in special cases.
‡$\Delta x \Delta p = h$ and $p = h/\lambda$; therefore, $\Delta p/p = \lambda/\Delta x$. Now, assuming that $\lambda \simeq 5 \times 10^{-7}$ m and $\Delta x \simeq 10$ m, we find $\Delta p/p \simeq 5 \times 10^{-8}$.

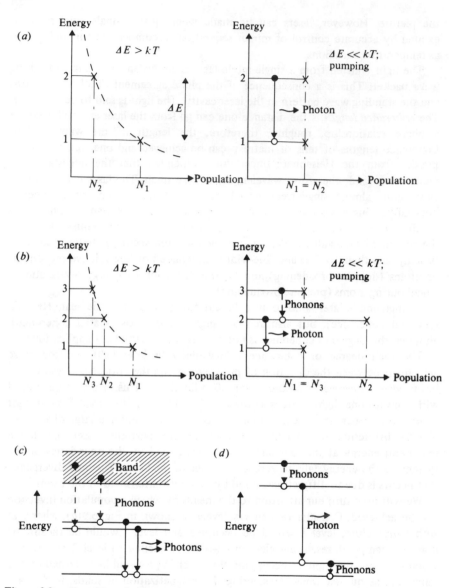

Figure 6.9 Transitions leading to laser action. (*a*) Two levels; (*b*) three levels; (*c*) excitation to a band, the ruby laser; (*d*) four levels, the neodymium glass laser.

where N_1 is the number occupying state 1 and ΔE is the energy difference between level 1 and level 2.

At normal temperatures ($\Delta E > kT$), N_2 is much less than N_1, but at very high temperatures, where $\Delta E \ll kT$, the population of the levels is almost the same.

The equivalent of very high temperatures can be achieved by providing energy to excite electrons by external means, such as an intense flash of light. This process is called 'pumping'. The best situation one can achieve with two levels is an equal population, but no amplification can be achieved with this because for every photon that stimulates an emission by decay of an electron from level 2 to level 1, another is absorbed to excite an electron from level 1 to level 2.

To achieve laser action, we need to have more electrons in the excited state than in the ground state. This situation can be achieved with a three-level system, such as that illustrated in Fig. 6.9(b). Electrons are pumped from level 1 to level 3 and an equal population in these two levels can be achieved. In the example shown, electrons decay from level 3 to level 2 by exchange of energy with phonons (a non-radiative decay)† and then from level 2 to level 1 by stimulated emission of photons. If the electron lifetime in level 3 is much less than in level 2, one can get a build-up of electrons in level 2 and a population inversion with respect to level 1. We have the necessary conditions for amplification and laser action.

The ruby maser has a three-level system. Ruby is crystalline alumina (Al_2O_3) with a few parts per million of chromium (Cr^{2+}) as an impurity. The localized chromium ground state splits into four levels in a magnetic field, and three of these are used to amplify by stimulated emission at *microwave* frequencies.

Ruby is also used for laser action, where the higher excited states of the chromium ion are used. The situation here is even more favourable for laser action. This is because the electrons are pumped into a *band* of states, so that all the light from the source of pumping light can be used, even if this emits a broad spectrum of wavelengths. A xenon flash tube is used for ruby lasers. The transitions are illustrated in Fig. 6.9(c). There is a non-radiative decay from the band to a finely split doublet intermediate state and stimulated emission takes place between this pair and the ground state, producing red light at wavelengths of 6929 Å and 6943 Å.

Another laser material that is frequently used is neodymium glass, which is a calcium tungstate glass with Nd^{3+} ions as an impurity. This has a four-level system, as shown in Fig. 6.9(d) and, in this case, as the laser action is between intermediate excited states, the population of the ground state is unimportant and it is even easier than for the three-level system to achieve laser action.

Laser action can be achieved in semiconductors and the principles are very simple.‡ A crystal of a direct-gap semiconductor like gallium arsenide is prepared with a junction between heavily doped p-type (hole conduction) and n-type (electron conduction) regions (called p^+ and n^+, the plus sign indicates that the regions are heavily doped).

---

† This will result in heating, with the result that solid-state lasers are usually only operated in a pulsed mode.

‡ As in many engineering situations, there are practical problems such as contacts, heat dissipation, and restriction of current pathways.

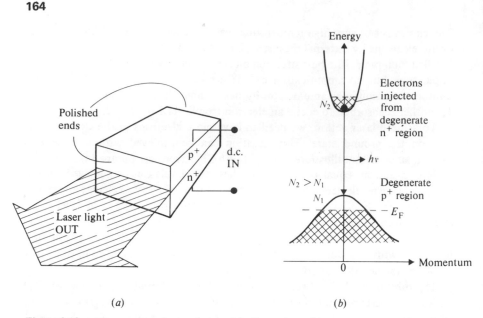

Figure 6.10 The semiconductor laser. (*a*) Geometry; (*b*) energy versus momentum diagram in p⁺ region close to the p–n junction.

Two opposite faces of the crystal normal to the p–n junction are polished and this forms the resonant cavity within the crystal (Fig. 6.10). The pumping is by a low d.c. voltage of polarity such that current flows across the junction (forward bias). This means that electrons are injected by current flow in the conduction band into p-type material, where the upper levels of the valence band are empty. We have a population inversion and, therefore, laser action and emission of light in the plane of the p–n junction. This is the same principle as is applied in normal (non-coherent) light-emitting diodes, but the geometry is quite different, as a resonant cavity is not used in the latter case. Gallium arsenide emits light in the near infra-red of wavelength 8383 Å (E_g = 1.48 eV).

The band gap can be increased by addition of a third component to make a ternary semiconductor such as gallium-aluminium-arsenide, which emits light at visible wavelengths.

6.5 The use of lasers — three case histories

There was a time when the laser was described as an invention looking for its mother — necessity. This is no longer true — the uses of lasers are many and varied. In some applications, such as surveying for civil engineering projects, use is made solely of the low divergence; the laser beam effectively becomes a long ruler. In other applications, the coherence achievable with single-mode operation is important,† whilst still further applications make use of the ability of lasers to produce very short but very high power pulses of light.

We will describe one example of the use of lasers from each of the three major application areas: for measurement, as a heat source, and as a probable future system of communications.

(a) Precise measurement of distance

Lasers are used for very precise mechanical positioning. For example, this is necessary for very accurate machine tools and in the production of masks for electronic integrated circuits. Use is made of the high coherence of a single-mode laser. A mirror which reflects the laser light is placed on the part to be measured. The reflected laser light is re-reflected by a system of mirrors and allowed to interfere with the light directly from the laser. A pattern of interference fringes representing the nodes in a standing wave pattern is produced. If the part to be measured moves by a distance of one fringe ($\lambda/2$), the illumination on a detector placed in the standing wave pattern changes from zero through a maximum to zero again. By counting fringes (electronically), it is possible to measure changes of length and the uncertainty in measurement can be as little as one tenth of a fringe. Thus, for the maximum distance where this technique is usable (the coherence length of the laser) of about 10 m, the accuracy of measurement is 1 part in 10^8.

The distance to the moon has also been measured using lasers, but by the very different 'pulse–echo' technique. This makes use of very short high-power pulses and the low divergence of laser light. The very short pulses are produced by a technique called Q switching. The amplification is suppressed during pumping, for example, by using a rotating end mirror, until complete population inversion is achieved. Then the cavity is 'switched-on', light is reflected back, amplification takes place, and all the stored energy is released in a burst of light which can last less than 10^{-9} s. Such a pulse was directed at the moon (through a telescope to further reduce divergence), reflected from an array put there during the Apollo 11 mission, and, 2.5 s after launching a pulse of 10^{20} photons, a total of 25 photons were detected as arriving back on Earth. Accurate measurement of the time taken enabled the distance between the earth and the moon to be measured with an uncertainty of ±15 cm.

The pulse–echo system is also used for detecting atmospheric pollution.

---

†One such application is holography. We have seen in Chapter 1 how the regular array of atoms in a crystal scatters electrons which interfere to create a diffraction pattern (Fig. 1.1.) Laser light scattered from different parts of a solid object can be made to create a diffraction pattern on the surface of a photographic plate. When the resulting photograph is illuminated with laser light, of the same wavelength as before, the original wave-front can be reconstructed and a three-dimensional image of the solid object can be made to appear.

(b) Lasers as a heat source

The very low divergence of laser light (achievable when there is good spatial coherence) and high spectral purity combine to present the possibility of using lenses to focus the beam to a diameter approaching the wavelength of the light. A beam that is a few millimetres in diameter can be focused to a few micrometres in diameter. A ruby laser with a power of 10 MW and a divergence of 10^{-2} radians can be focused to produce a surface power of 10^{13} W m^{-2}, five orders of magnitude greater than the most powerful of conventional welding techniques. This high power will result in evaporation or sublimation of the surface and is a good way of drilling holes† and cutting materials. For example, lasers are used to trim microcircuit resistors by cutting very small slots which restrict current pathways and, therefore, increase the value of resistance.

To produce a very fine weld, the power and pulse length have to be carefully controlled so that the material is melted but not evaporated. Materials as varied as polythene, steel, tantalum, and quartz have been welded using lasers.

The short exposure time, small spot size, and ease with which the equipment can be handled (using glass fibres as light guides) has led to the use of lasers for eye surgery, where a detached retina can be re-attached to the back of the eye by coagulation (caused by local heating) without damage to the rest of the eye.

(c) Laser communications

Information can be transmitted by photons from one place to another. This is done by using relatively high-frequency photons (called the carrier) and modifying the properties of the carrier, such as amplitude or frequency, at a lower frequency (called the modulation frequency). The modulation frequency carries the information we require and is usually about one tenth of the carrier frequency. A photon of $\nu = 10^6$ Hz is in the medium-wave region and about 25 speech channels can be carried between 0.3 and 3 MHz (the medium-wave band) without overlap of frequencies (we need a channel width of at least twice the modulation frequency). In the UHF band (0.3×10^9 to 3×10^9 Hz), ten thousand speech channels or ten TV channels can be transmitted. If we could modulate photons in the visible spectrum (10^{13} to 10^{15} Hz) at only 0.1 per cent of the photon frequency, the optical band could carry up to 10^5 TV channels or 10^8 speech channels without any overlap. The growth in world communications is so rapid that there will soon be a need for this kind of capacity, so the race is on to develop a practical system of communications at optical frequencies. This has only become possible with the availability of a source of light of high coherence and low divergence — the laser.

Laser light can be modulated by selecting a plane of polarization of the light using a polarizer. If the light is then passed through an electro-optic crystal (such as a ferroelectric crystal), the plane of polarization can be rotated by the

† These can be very deep due to internal reflection inside the hole.

field-induced polarization in the crystal. A varying electrical signal supplied to the polarizer varies the plane of polarization of the emerging laser light, which affects its ability to pass through a second polarizer. The light has become intensity modulated. High power is needed to drive electro-optic cells at high frequencies because of the large amount of stored energy (capacitance).

An alternative approach is to use pulse code modulation (a sophisticated form of Morse code), where the information is transmitted in digital form (is there a pulse or is one missing?). Lasers can readily be induced to emit a train of very narrow pulses by interference of the axial modes (beats). Electro-optical crystals are used to carry out the modulation by elimination of selected pulses.

The transmission of laser light would not usually be through the air, because of absorption and changes in refractive index; hence, light guides will have to be developed. The best method would appear to be the use of glass fibres, where total internal reflection guides the light, although losses are rather high at present. Optical integrated circuits using semiconductor lasers as sources with light guides in the surface, electro-optic components, semiconductor detectors, and fibre-optical connectors are visualized as future developments.

Intensity modulation of laser light is already in use for distance measurement. The change in phase of the modulation frequency after reflection from a target is dependent on the distance the wave has travelled. The modulation frequency can be high, so a resolution of 1 mm at 1 km range can be achieved. For example, the bulge in the wall of a dam under the weight of water can be measured. Also, because of low divergence, a small target area can be selected, so the profile of small objects on the ground can be detected by a flying aircraft.

Part 4

Materials

'Of sooty coal th'empiric alchymist
Can turn, or holds it possible to turn
Metals of drossiest ore to perfect gold.'

('Paradise Lost' by John Milton)

7. Metals

7.1 Introduction

Up to this point, we have been concerned mainly with the atomic scale of things. The properties of solids such as bonding, band structure, and magnetic and dipole moments have been described in terms of the electron distribution around an atom or small group of atoms. The crystal structure is considered to be a perfect reproduction of this unit throughout space. Defects in the crystal structure will complicate the picture (for example, we have seen how crystal defects will reduce thermal and electrical conductivity at low temperatures), but it has not been necessary to consider crystal defects in detail in order to understand the phenomena that have been described so far.

We now come to consideration of mechanical properties of solids such as strength, toughness, and ductility. Also, we will consider how those materials such as steel, brick, concrete, and glass are made and formed into the required shapes. Mechanical properties of solids depend sensitively on the nature of defects in the crystal structure and the way crystals are combined in the shaped article. Our attention has now to be turned to things bigger in scale than hitherto.

Many features larger than the unit cell have to be understood before we can describe how to produce a material with an optimum set of properties for a particular application. This is the province of materials science or materials engineering. This last part of the book is an introduction to the subject. Once more, we have avoided mathematics, concentrated on the principles, and extracted simplified examples from the complication of fine detail. There is a lot of fine detail in materials science; for example, the number of metal alloys one can make are infinite. The number of variations on treatments of just one alloy, for example, the carbon—iron system (steel and cast iron), seems almost infinite. These chapters, therefore, are not intended as an exhaustive guide to materials engineering, but just as a taste of the concepts, methods, and classes of materials that the materials engineer uses to shape our world. The section is divided into three chapters: the first two, metals and non-metals, deal with the principles; the third gives case histories of materials engineering using the principles established and describing production techniques.

7.2   Defects in solids

One of the reasons why metals are crucial to technical advancement is that their mechanical properties can be varied between the high strength of tool steel and the ductility (or plasticity) of beaten gold. Ceramics possess high strength but no ductility, whilst plastics have ductility but little strength. Only in metals can these two important properties be combined and their balance controlled by the materials engineer.

This balance is controlled by modifying the type and distribution of defects in the crystal structure. These are introduced below, categorized by their dimension in space.

(a)   Zero dimensions − point defects

This is the type of defect that we have already encountered in the last two chapters when discussing their effect on phonons and photons in solids. Three types of point defect are illustrated in Fig. 7.1. The vacancy is just the absence

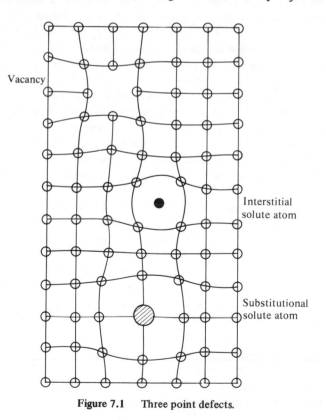

Figure 7.1   Three point defects.

of an atom in the crystal structure. Vacancies are more plentiful at high temperatures, where the phonons are of sufficient energy to displace atoms from their sites. This has the effect of increasing the volume of the solid, adding to the thermal expansion from anharmonic phonon vibrations. Vacancies are very important in that they make it possible for atoms to move through the crystal of a solid. An atom neighbouring the vacancy can jump into the vacant site, aided by energy supplied by phonons, because the potential barrier is low in this direction. The vacancy will have effectively moved by one lattice spacing to the site of the atom that has just jumped. By a succession of hops like this, the vacancy can migrate large distances, and atoms can migrate in the opposite direction by hops into vacancies.

The interstitial is an extra atom fitted into a potential well in the crystal structure that is normally not occupied by the host atoms; for example, the body centre in a face-centred cubic structure. For an interstitial to fit into the structure easily, it has to be very small; thus, atoms in the first row of the periodic table such as carbon and nitrogen are the normal type of interstitial defect. This is called an interstitial solid solution if the small atoms can be introduced without altering the crystal structure (although they may produce a large strain).

If an interstitial is created by displacing one of the host atoms using energetic radiation, then the strain on the crystal structure is very great and the interstitial may diffuse rapidly to the site of a nearby vacancy and be eliminated.

A third type of defect is the foreign atom that is sufficiently similar in size and electronic configuration to the host atoms that it can substitute for a host atom in the crystal structure. We have met this behaviour when discussing doping of semiconductors in Sec. 3.13. This is also an important feature of alloys; it is called a substitutional solid solution. Most metals will tolerate a small amount of another metal in solid solution at high temperatures, but for some alloys the atomic sizes and crystal structures are sufficiently alike for substitutional solid solutions to form at any composition between 100 per cent of one constituent and 100 per cent of the other. This continuous series of solid solutions occurs, for example, with silver—gold alloys and copper—nickel alloys.

The role of point defects, especially substitutional atoms, in increasing the strength of metals will be dealt with later. One aspect that can be mentioned here is diffusion.

We have mentioned that lattice vacancies can migrate through the crystal. This is called diffusion, and it means that any non-uniformity, for example, in the distribution of foreign atoms in solution, will tend to even out (if the temperature is high enough). This occurs by means of the migration of substitutional atoms travelling in the opposite direction to the vacancies and using the vacancies as sites for each jump, or by direct diffusion of interstitials from one interstitial site to neighbouring ones. Diffusion is a purely random process which can be seen in operation when a puff of smoke is blown into the

Concentration of
impurity (log scale)

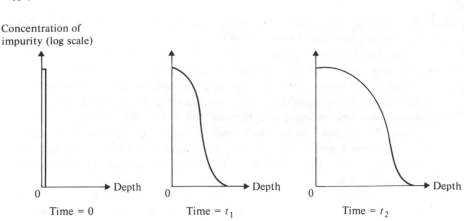

Figure 7.2   Diffusion of an impurity into the surface of a solid.

corner of a room — eventually, the room is filled uniformly with smoke due to the random motion of the smoke particles. The concentration of impurity with time below the surface of a solid which is in contact with a source of foreign atoms is shown diagrammatically in Fig. 7.2.

Diffusion of this type is used in the surface treatment of steel gearwheels where carbon is diffused into the surface (carburizing), resulting in a very hard, wear resistant surface, whilst the bulk of the metal retains its ductility and so is resistant to brittle fracture. Diffusion can also be important in controlling the rate at which one crystal structure transforms into another when a metal is cooled.

(b) One dimension — line defects

The secret of the ductility of metals is that one plane can easily slip across another plane when the crystal is under mechanical stress. The slip does not occur instantaneously across the whole plane. This would need too much energy, as we would have to simultaneously break millions of atomic bonds. Slip occurs progressively by introduction of a discontinuity in the form of a linear defect in the crystal structure. This is called a *dislocation* One type of dislocation is illustrated in Fig. 7.3(*a*). This will be described in detail later, but we can say here that the dislocation is a region crystal disorder marking the boundary between the unslipped perfect crystal and the slipped crystal, which, as it has slipped by one lattice unit, is also perfect. The dislocation glides through the crystal, unzipping a row of atoms at a time and, by small atomic rearrangements, zipping them up again displaced by one lattice spacing. The dislocations are the key to the ductility of metals, and restriction of dislocation movement is the key to control the strength of metals. All the other forms of defect that we are discussing in this section have some role to play in influencing dislocation movement.

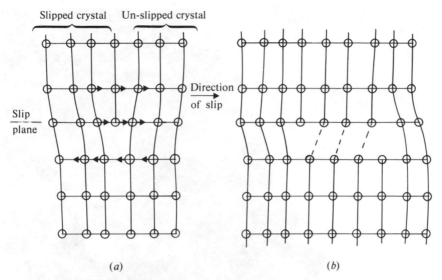

Figure 7.3 Two crystal defects. (*a*) An edge dislocation (arrows indicate atomic movements close to the dislocation); (*b*) a stacking fault.

(c) Two dimensions — area defects

We mentioned in Sec. 2.7, when discussing the close-packed structures, that sometimes we do not get regular packing (such as *ABCABC* for f.c.c.) throughout the close-packed structures. Stacking faults do occur in high-symmetry structures. The face-centred cubic metals copper and gold and the face-centred cubic alloys brass and stainless steel are good examples. The normal form of a dislocation in these materials includes an area of stacking fault. The dislocation (Fig. 7.3(*b*)) takes the form of a ribbon of stacking fault bounded by partial dislocations. The first one unzipping the structure by part of a lattice spacing to create the stacking fault and the second one finishing the job by zipping up perfect crystal, once more slipped by one lattice spacing. The ribbon of stacking fault is widest for stainless steel, where it can extend for 50 lattice spacings. The movement of this type of dislocation is more easily disrupted, which gives metals with stacking faults very rapid work hardening compared with metals with simple line dislocations.

(d) Three dimensions — volume defects

When a metal grows from the melt, many crystals start forming at the same time. There is usually no correlation between the orientation of these different crystals; thus, when the whole mass is solid, there is an abrupt transition from one orientation to another where the crystals meet. The polycrystalline form of metals shows dramatically at the surface when the metal is grown from vapour as is illustrated in Fig. 7.4. The crystals are called grains and the structure, which is

Figure 7.4 Very pure vapour-deposited chromium, showing crystallites at the metal surface.

the normal structure of metals that have been solidified quickly, is poly-crystalline. Dislocations cannot travel easily across the grain boundaries, so the latter have a role to play in the control of mechanical properties of metals. In addition to this, the distortion of the crystal structure at the grain boundaries provides gaps, which are favourable sites for point defects to sit. Point defects, therefore, tend to diffuse to these 'sinks', forming clusters.

If the metal is an alloy of two or more constituents which do not form a continuous series of solid solutions, then two or more types of grain with completely different composition and crystal structure will form. This so-called multi-phase structure places even more restrictions on dislocation movement. Control of the nature and distribution of the phases, as the different types of grain are called, is a very important part of the materials engineering of alloys such as steel and Dural as we will discover later.

Two other types of volume defect which are always undesirable (unlike the defects discussed so far) are porosity and inclusions. Porosity arises from gas bubbles evolved during casting. Shrinkage on cooling can also leave cavities. Inclusions are lumps of foreign matter that are cast, rolled, or drawn into the metal. We depend on the design of moulds, the skill of the foundryman, and the cleanliness of the mill operators to eliminate these problems.

In the rest of this chapter, we will describe the solidification of elemental metals and alloys from the molten state as well as the grain structures and phase structures that are formed. We will then describe the mechanical properties of metals and explain the dominating influence of dislocations and how they operate. The dislocation interactions that lead to the hardening of metals will be itemized, and the softening of metals by annealing will also be described. The chapter finishes with a brief description of the failure of metals by fracture, creep, and fatigue.

7.3 Growth of solids from the melt

At high temperatures, where the thermal energy is of the order of the interatomic bond strength, materials have the loose random structure and rapid diffusion rate characteristic of a liquid. Below a certain temperature, energy is saved by formation of a regular crystal structure. Given the right conditions, the material will solidify and the energy saved is given out as heat (the latent heat).

The solid starts to form when enough atoms congregate together in a regular array to create a stable nucleus.

The radius of the smallest stable nucleus, r^*, is defined as the radius (assuming for simplicity that the nucleus is spherical) at which the energy saved on creation of a regular crystal structure (the energy saved proportional to volume $- r^3$) becomes greater than the energy used to create a surface (which is proportional to surface area $- r^2$). This is illustrated in Fig. 7.5. The critical radius is infinite at the melting point, so we have to cool below the melting point (supercooling) before solid will start to form. If we cool to just below the melting point, then there is only a small energy saving on creating a solid crystal structure and so the size of the critical nucleus is very large. Although it is improbable that a large number of atoms arrive together at the same time, the rate of diffusion is very rapid and so some nucleation will take place. The nucleation density will be low and so, as each grain may grow from a single nucleus, the casting will have a large (coarse) grain size.

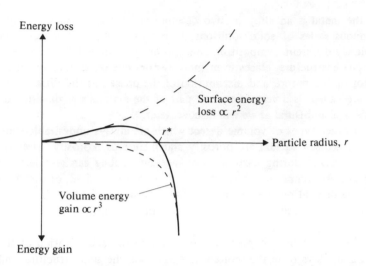

Figure 7.5 The energy balance as a function of the size of a solid nucleus. $r*$ is the radius of the critical nucleus.

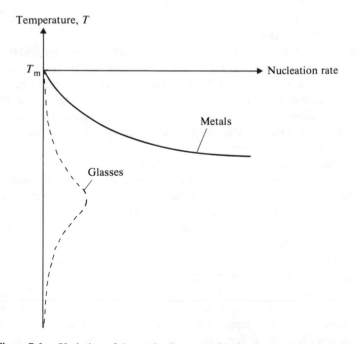

Figure 7.6 Variation of the nucleation rate with the degree of supercooling.

If we cool rapidly to achieve a high degree of supercooling, then the size of the critical nucleus drops sharply and, although there is a drop in the rate of diffusion, the overall nucleation rate is much higher than the case for slow cooling; we obtain a fine-grained casting.

For metals, because of the non-directional bonding, diffusion is fairly rapid even well below the melting point. This is not the case for the covalently bonded solids — especially the oxides. Diffusion is so sluggish that, if one cools rapidly, a temperature can be reached at which no nuclei can form, even though the size of the critical nucleus is very small (because effectively no diffusion can take place). If nucleation is suppressed entirely, then the liquid structure is retained at low temperatures; we have a glass. The effect of supercooling on nucleation rate is shown in Fig. 7.6.

We can almost never cool metals rapidly enough to suppress nucleation (although some research workers may have succeeded by very very rapid cooling). Certainly, in any practical situation, nucleation of the solid starts not in the bulk of the liquid by random chance (homogeneous nucleation) but at irregularities on the mould surface, on dust particles, or any other convenient surface (heterogeneous nucleation). Sometimes, particles are deliberately added to the melt to increase the nucleation rate and achieve a fine-grained structure. This is done for aluminium—silicon alloys used to make car engine blocks.

The variation of temperature in a casting as a function of time is shown in Fig. 7.7. Immediately after nucleation, the crystal growth rate is rapid because of the degree of super-cooling; but latent heat is released and, as a result, the temperature rises to the melting point and stays at the melting point until solidification is completed. The rate of crystal growth subsequent to nucleation depends on the rate at which heat is removed from the mould.

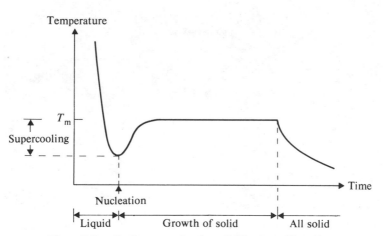

Figure 7.7   Cooling curve for the solidification of a casting.

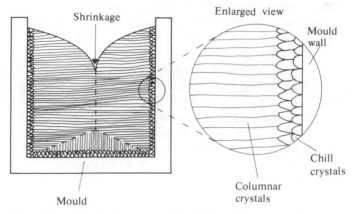

Figure 7.8   Schematic representation of the grain structure of a casting.

A simplified picture of the structure of a normal casting formed when molten metal is poured into a relatively cold mould is shown in Fig. 7.8. At the mould walls, there is a rapid rate of cooling and, therefore, a high nucleation rate. Small grains (chill crystals) are formed. The interior of the casting cools more slowly, the crystal growth rate is slow, and large columnar crystals grow inwards from favourably oriented chill crystals to meet at the centre of the casting.

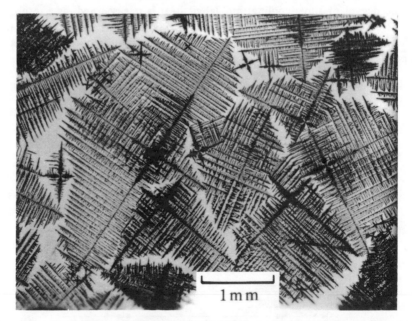

Figure 7.9   Dendrites in solidified ammonium chloride (easier to see than in metals, where neighbouring crystals interfere).

The reduction in metal volume due to the reduction of interatomic spacing when the change of state from liquid to solid occurs creates a cavity on the top surface of the casting. This shrinkage has to be allowed for in the design of the mould, where sometimes the top has a smaller section 'hot header' which contains all the shrinkage. After solidification, this is cut off the ingot and reused in the next melt.

In two-phase alloys, it is often the case that one phase solidifies first, rejecting atoms of the second phase into the melt. This process is called *segregation*.

In many cases, the crystal structure is such that the grains grow much faster in certain crystal directions. When we have a two-phase alloy, this preference for growth in certain directions can be satisfied because the crystals grow in the melt due to segregation and, therefore, there are no other crystals in close proximity. The crystals form branches and from these branches other branches grow until we have the beautiful structures illustrated in Fig. 7.9. This is called dendritic (tree-like) growth. The snowflake is an example of this type of growth.

7.4 Solidification of alloys – phase diagrams

The way that an alloy solidifies is more complex than an elemental metal because we have two or more constituent atoms. We will confine ourselves to alloys with just two constituents (binary alloys) as these are complicated enough and all the principles established can be applied to alloys with more constituents. We will also ignore pressure as a variable, assuming that it is always constant and at standard atmospheric pressure. Finally, we will assume that, in the liquid state, both metals mix completely, they are *miscible* (unlike oil and water).

If we have an alloy of A and B, then, at any one temperature, we may have in thermal equilibrium any one or more of the following phases: liquid (A B), solid A, Solid (A B) (there may be several different combinations), intermetallic compound ($A_x B_y$) (perhaps more than one), and/or solid B. The phases that are stable are determined by atomic bonding and the way this is displayed (in a series of large books that the materials engineer can look up) is in the form of a *phase diagram*. This is a graph of temperature against composition divided into areas representing the range of composition and temperature over which a particular phase or mixture of phases is stable. Some phase changes are rather sluggish, however, especially in the solid state, so one can beat the equilibrium phase diagram by cooling quickly (quenching) and freeze-in the higher-temperature phase structure. This is important for the materials engineer and examples of the use of quenching will be given in the last chapter. In this section, we will describe four types of phase diagram using real alloys as examples. These examples cover most types of phase transition that exist.

(a) A constant series of solid solutions

This is the simplest of phase diagrams because the solid is a single phase with both types of atom included in the crystal structure; hence, we have the transformation

$$\text{Liquid (A B)} \xrightarrow{\text{cooling}} \text{Liquid (A B)} + \text{Solid (A B)} \xrightarrow{\text{cooling}} \text{Solid (A B)}$$

or

$$\text{L} \rightarrow \text{L} + \alpha \rightarrow \alpha$$

where L represents the liquid phase and α represents the solid phase,† (a mixture of A and B.)

We will take as our example the Cu–Ni alloys. The appropriate phase diagram is shown in Fig. 7.10. It seems strange at first that liquid and solid can exist in equilibrium with each other. We can see why this occurs if we consider what happens when we rapidly cool a molten mixture of copper and nickel, say of composition C_I, to a temperature T between the melting points (T_m) of copper and nickel, and allow time for thermal equilibrium to be established. If the temperature is low enough, nucleation of the solid phase will take place. The solid phase will tend to be richer in nickel because more energy is saved in making nickel–nickel bonds than copper–nickel or copper–copper bonds, and so nickel–nickel bonds have a slightly higher probability of forming. That nickel

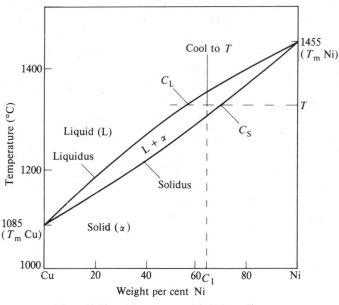

Figure 7.10   The copper–nickel phase diagram.

---

†Greek letters α, β, γ, δ and so on, are used to label the solid phases starting with the first on the left.

has a higher bond strength than copper is indicated by the higher melting point. The solid will therefore have a composition shown as C_S (richer in nickel than C_I); this solid is in thermal equilibrium with a liquid that, as a consequence of this, is richer in copper, with composition C_L. The phase diagram, therefore, is composed of two lines, indicating the temperatures where the two transitions given in the equation above occur. The upper line, above which all is liquid, is called the liquidus, the lower line, below which all is solid, is called the solidus. In between the liquidus and solidus we have a nickel-rich solid in equilibrium with a copper-rich liquid.

In a real situation, the alloy is not held at a constant temperature but is cooled continually. If the cooling is slow, then the first solid to form will be almost pure nickel. We, therefore, get segregation within the ingot: the material which is first to solidify, i.e., the material near the mould walls, is rich in nickel; the centre of the ingot is rich in copper. This lack of homogeneity in the ingot can be removed by subsequent working to break up the cast structure and, by heat treatment, at temperatures that are below the solidus but sufficiently high to allow diffusion, to thoroughly mix the constituents.

(b) Very restricted solid solubility — the eutectics

This is the opposite situation to that just described, in that the bonding and crystal structure of the two components are incompatible. Hence, only a very small amount of one constituent can fit into the crystal structure of the other. We have very limited solid solubility. The transformation which occurs is

$$L \xrightarrow{\text{cooling}} L + \alpha \text{ (or } L + \beta) \xrightarrow{\text{cooling}} \alpha + \beta$$

Where α is almost pure A and β is almost pure B.

We will use as our example an alloy between two elements from our pet row of the periodic table: aluminium and silicon. The phase diagram for Al–Si alloys is shown in Fig. 7.11. A binary alloy of this type is very common and is called a *eutectic* (meaning 'of low melting point'). The α phase is almost pure aluminium (just a very small amount of silicon, less than one per cent, in solution) and the β phase is almost pure silicon.

If we add a small amount of silicon (say five per cent) to molten aluminium, we find that it dissolves completely (like sugar in tea), even if the aluminium is only just above its melting point (660 °C). This is perhaps surprising, as the silicon melting point is 1430 °C, but there are so few silicon atoms that a stable silicon nucleus cannot form.

Solid silicon is unstable in the melt at this temperature. Silicon atoms will continually leave the surface and not be replaced because they are so low in concentration in the melt. Eventually, all the silicon will have gone into solution.

In fact, we have to cool some way below the aluminium melting point (to temperature T_1) before any solid (almost pure Al) can form. This is because the presence of silicon atoms reduces the chance of an all-aluminium nucleus

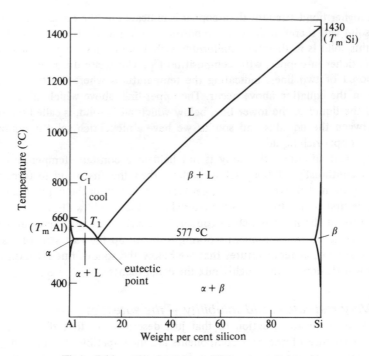

Figure 7.11   The aluminium–silicon phase diagram.

forming; a mixed nucleus is impossible because the aluminium can only take a very small amount of silicon in solid solution. As we cool below T_1, more and more solid aluminium is formed and the liquid becomes more rich in silicon until the liquid reaches 11.6 weight per cent Si at 577 °C. At this point, it solidifies completely into a mixture of almost pure Al (α phase) and almost pure Si (β phase). This point on the phase diagram (11.6 weight per cent Si, 577 °C) is called the eutectic point. At this composition, there is enough silicon in solution for a stable silicon nucleus to form and the alloy freezes to form a very fine mixture of the two phases.

Similar arguments apply to the silicon-rich side of the eutectic point, so we can construct the whole alumium–silicon phase diagram as shown.

Alloys cast from a melt of eutectic composition are composed of a very fine mixture of the two phases. Alloys to the alumium-rich side of the eutectic point have large dendrites of Al in a eutectic matrix. Alloys to the silicon-rich side have large blade-like crystals of silicon in a eutectic matrix. These silicon crystals make the alloys difficult to machine and leave a rough finish. This problem can be overcome by the addition of phosphorus. This leads to creation of aluminium phosphide crystals, these act as nuclei for the silicon phase which takes the form of very many small silicon crystals; on solidification this results in a very refined grain structure.

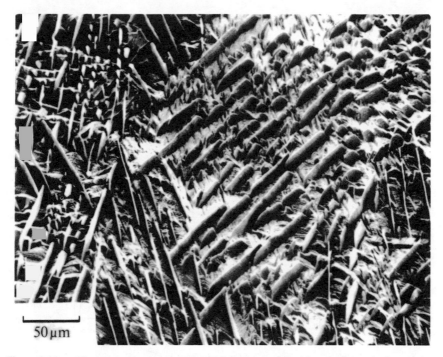

50 µm

Figure 7.12   Silver-rich plates in a silver–bismuth eutectic (chemically etched, scanning electron microscope).

An electron micrograph of the surface of a silver–bismuth eutectic is shown in Fig. 7.12.

(c) Intermetallic compounds

Sometimes, the two components of an alloy at a particular composition form a regular crystal lattice in which each component has its appointed place. This is called an intermetallic compound and the example we will give is in the Mg–Pb alloy system, where the compound Mg_2Pb is formed at 33 atomic per cent lead.† The phase diagram is shown in Fig. 7.13. The intermetallic compound is not given a Greek letter, but is named by its composition. This diagram is really two eutectics joined together. One between Mg and Mg_2Pb and the other between Mg_2Pb and Pb.

An important intermetallic compound that we will meet later is iron carbide (Fe_3C), which occurs at 25 atomic per cent or 6.7 weight per cent carbon.

---

†Usually, phase diagrams are plotted as weight per cent rather than atomic per cent as this is more practical if one has to weigh out the constituents to make an alloy. Often, there is not much difference, if the weights are similar. In this case, however, there is a very great difference in atomic weight and the compound forms at approximately 80 weight per cent Pb.

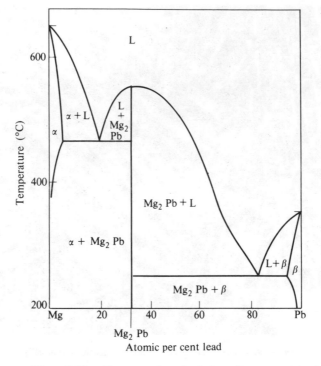

Figure 7.13 The magnesium–lead phase diagram.

(d) More complex phase diagrams

The Mg–Pb phase diagram looks very complicated, but it can be simplified, as we have said, by considering it as two eutectics back-to-back. We will now look at an even more complex phase diagram, that for Cu–Zn alloys (Fig. 7.14). This is one of the more complex diagrams. There are six phases going from α (Cu with Zn in solid solution) to η (Zn with a little copper in solid solution). The best known of the Cu–Zn alloys is brass, which is usually made from alloys within the α-phase field. These are, therefore, solid solution alloys similar to Cu–Ni, but some of the higher zinc brasses have some β phase in them. The purpose of showing you such a complex phase diagram is not to go into the intimate details of the system, but to draw some inferences about phase diagrams in general and to show examples of two types of phase transformation not yet encountered.

The two transformations are indicated as dotted lines on Fig. 7.14(a) and are shown enlarged in Fig. 7.14(b) and (c). The first is very similar to the eutectic transformation, except that all the phases are solid, and, in this instance,

$$\delta \xrightarrow{\text{cooling}} \delta + \gamma \;(\text{or } \delta + \epsilon) \xrightarrow{\text{cooling}} \gamma + \epsilon.$$

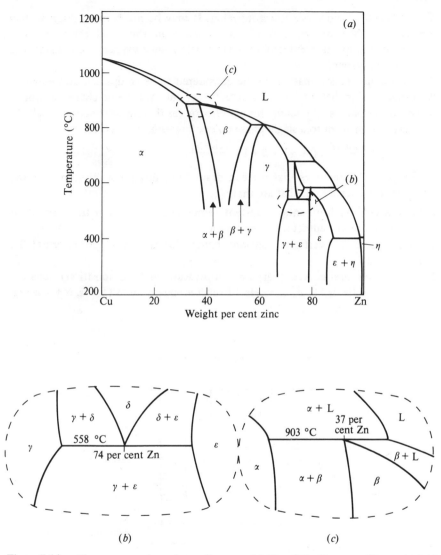

Figure 7.14 The copper–zinc phase diagram. (a) The full diagram; (b) a eutectoid transformation; (c) a peritectic transformation.

This is called an *eutectoid* transformation. It may be much more sluggish than the eutectic because we depend on diffusion in the solid state. We will encounter an important eutectoid transformation when we come to consider the iron–carbon system.

The second transformation can be considered to be an upside-down eutectic. We mean by this that whilst, for a eutectic (or eutectoid), one phase transforms to two other phases on cooling, for a *peritectic* (as this transformation is called), two phases transform to a single phase on cooling; thus, in this case,

$$\alpha + L \xrightarrow{\text{cooling}} \beta + \alpha \ (\text{or} \ \beta + L) \xrightarrow{\text{cooling}} \beta.$$

The general inferences that we can draw from this phase diagram and the three others we have looked at are that:

1. lines which are horizontal (constant temperature) denote a transformation reaction (eutectic, eutectoid, or peritectic);
2. lines which are vertical (constant composition) denote an intermetallic compound;
3. lines which are curved (varying temperature and composition) separate single-phase fields (L, α, β, γ, etc.) from two-phase fields (L + α, $\alpha + \beta$, $\beta + \gamma$, etc.).†

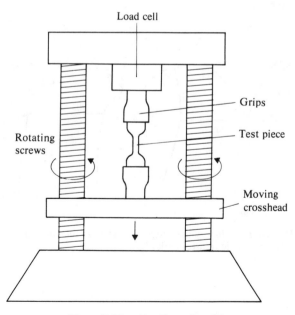

Figure 7.15   Tensile testing rig.

†It is thermodynamically impossible for three phases to exist in equilibrium over a range of temperature *and* composition.

7.5 The mechanical properties of metals

(a) The tensile test

The mechanical properties of a metal are usually assessed in a tensile test. In this test, a specimen in the form of a rod (of round or rectangular cross-section) is clamped at the ends (Fig. 7.15). The separation of the clamps (or grips) is then steadily increased and the load on the specimen is measured as a function of the extension of the specimen.

This is converted into a plot of stress against strain, where

Engineering stress, σ = Load ÷ Original cross-sectional area,

Engineering strain, ϵ = Change in length ÷ Original length.†

An idealized (engineering) stress–strain relationship is shown in Fig. 7.16 in order to illustrate the important features that are used to specify the properties of a metal.

For simplicity, the curve can be divided into elastic deformation and plastic deformation (although some plastic deformation takes place even at very low values of stress).

In the elastic region, the metal is stretched by increasing the length of atomic bonds. When the load is released, the bonds relax and the specimen recovers its original shape. The relationship between stress and strain is linear and the slope of the line is called the modulus of elasticity of the material (Young's modulus).

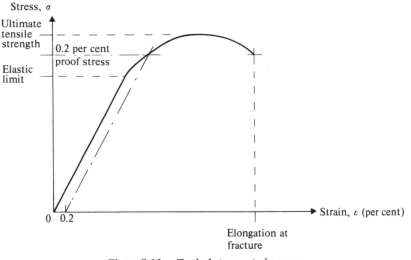

Figure 7.16   Typical stress–strain curve.

†To calculate true stress and strain for any particular loading, we have to use the area at this load and the incremental relative change in length as this load.

This is dependent on the atomic bond strengths and crystal structure of the metal.

At some loading, atomic bonds start to be broken in large numbers and, thus, when the load is released, the specimen does not recover its original shape. The load at which this happens is called the *elastic limit*.† At loads above the elastic limit, plastic deformation becomes the dominant process. As plastic deformation can take over gradually, the value of stress at the elastic limit is sometimes difficult to determine.

For this reason, the datum that is used in specifications of metal properties is sometimes the stress at which a line cuts the stress—strain curve when drawn parallel to the elastic region but offset by a certain amount of strain (0.2 per cent in this example). This is called the 0.2 per cent *proof stress* (0.2 per cent offset *yield strength*). Other data that are quoted in specifications are the maximum stress on the stress—strain curve (called the *ultimate tensile strength*), the percentage strain (or elongation) at fracture and the percentage reduction of area at fracture.‡ These last two figures are a measure of the ductility of a metal. The more ductile a metal is, the more it can be stretched (and the more it thins) before it breaks.

There is little the materials engineer can do about the elastic region except choose a material with bonding that gives the required modulus. He can alter the grain structure either to take advantage of the 'strong' directions in the crystal structure (as is done for the single crystal tungsten turbine blades mentioned in Sec. 2.14(c)) or to give uniform properties in all directions.

We will now consider plastic deformation, where atomic bonds are broken and irreversable changes in dimensions occur.

(b) Plastic deformation

If we take a single crystal of zinc, of a favourable orientation, and load it to beyond the elastic limit, lines appear around the specimen (often at 45° to the load axis). As the stress increases, the lines become sharper at it becomes clear that blocks of crystal are sliding across one another until fracture occurs, leaving a mirror-smooth plane across the specimen (Fig. 7.17). The plane along which the slip occurs is always a close-packed plane in the crystal structure. Thus, plastic deformation takes place by a slip of one close-packed plane of atoms across a neighbouring plane.

Zinc has the hexagonal close-packed structure and the close-packed planes are the base planes. There is only one plane of this type per unit cell (Fig. 7.18(a)), so the slip is simple and can be observed in the experiment just described if the {100} planes are at about 45° to the load axis.

† Some metals exhibit a well defined knee or kink in the stress-strain curve just above the elastic limit which is called the *yield point* (σ_Y).
‡ The 'turn-over' of the engineering stress—strain curve is due to the fact that we are using the original dimensions to calculate stress and strain. The true stress—strain curve does not exhibit this reduction in stress at high strains.

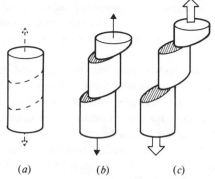

(a)　　　　　(b)　　　　　(c)

Figure 7.17 The effect of stress on a single crystal of zinc. (*a*) Just above the yield point; (*b*) higher stress; (*c*) close to the fracture.

The existence of only one set of slip planes for hexagonal metals limits the amount of deformation that can take place in a polycrystalline sample. Only very few grains will be oriented favourably for slip, and if no slip can occur brittle fracture will take place; these metals are, therefore, brittle.

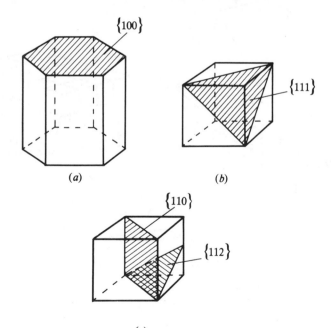

Figure 7.18 Slip planes in metal single crystals. (*a*) Close-packed hexagonal; (*b*) face-centred cubic; (*c*) body-centred cubic.

The close-packed direction for face-centred cubic crystals is the {111} plane (Fig. 7.18(*b*)) and eight planes of the {111} type exist in the unit cell. Thus, slip can occur on many planes and so the face-centred cubic metals are usually ductile.

The body-centred cubic metals come between f.c.c. and h.c.p. with slip planes of the type {110} and {112}, as shown in Fig. 7.18(*c*).

Now that we know that normally metals deform by slip, we can try to calculate the force necessary for this to occur. If we assume that slip occurs by simultaneously breaking all the bonds between atoms above and below the slip plane, then this calculation would predict that the force necessary to do this would be about 20 000 MN m^{-2}. But the strongest steel we can produce has a strength of about 3000 MN m^{-2}. Therefore, our assumption of simultaneous slip across the whole slip plane must be wrong.

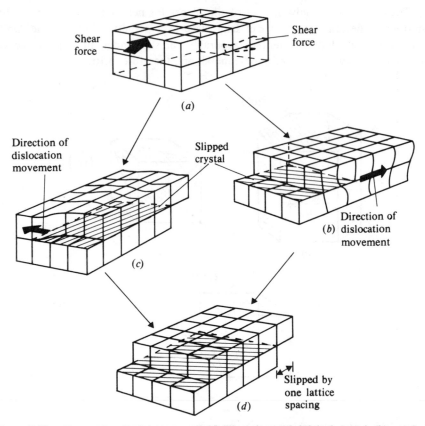

Figure 7.19   Progressive slip by movement of dislocations. (*a*) Original crystal; (*b*) an edge dislocation; (*c*) a screw dislocation; (*d*) crystal slipped by one lattice spacing.

In fact, as we have already stated in Sec. 7.2, slip takes place progressively by means of the movement of dislocations which are linear regions of distorted crystal structure marking the boundary between the slipped and the unslipped crystal. Dislocations come in two varieties which move in different ways to give the same end result.

This is illustrated in Fig. 7.19, where stress is applied to a perfect crystal. One way that this is absorbed is by bending the crystal structure and pushing an extra plane of atoms along in the direction of stress, as shown in Fig. 7.19(b). This type of dislocation is called an edge dislocation; it moves in the direction of the applied stress, unzipping one row of atoms at a time, until it runs out of a surface and the whole block of crystal has slipped by one lattice spacing in the direction of stress (Fig. 7.19(d)). The alternative method of causing slip is by unzipping a row of atoms (nearest to us in illustration (Fig. 7.19(c))) parallel to the direction of stress and slipping them by one lattice unit in the direction of stress. The dislocation, again, is the region of strained crystal between the slipped and unslipped regions; in this case, however, it lies parallel to the direction of stress and moves perpendicular to the direction of stress. The end result is the same as the case for the edge dislocation: the whole block will have slipped one lattice unit in the direction of the applied stress (Fig. 7.19(d)).

This second type of dislocation is called a screw dislocation, because if one traces a path around the dislocation from atom to atom along bonds in a plane perpendicular to the dislocation you do not get back to where you started: your path traces out a helix. Most beautiful spiral growths have been seen on the surface of crystals as a result of growth on a face intersected by screw dislocations.

Dislocations do not have to be purely one type or the other, they can be partly screwy and partly edgy; in fact, they usually take the form of loops within the crystal. These loops grow in diameter when stress is applied. Figure 7.20 shows schematically part of a mixed dislocation loop.

The dislocation, therefore, is the main cause of plastic deformation in metals. The more slip planes that exist in the crystal, the more ductile the metal will be. The more difficult that dislocation movement becomes, the stronger the metal will be; however, some ductility will have to be sacrificed. The different methods of hindering dislocation movement will be described in the next section.

Before proceeding to a description of the hardening of metals, mention must be made of a mechanism of deformation that does not involve slip. This is called twinning. The process is illustrated in Fig. 7.21. There is a shearing motion of atoms on one side of the twin boundary. The crystal structure of the twin so formed is a mirror image of the untwinned crystal. The atomic displacements during twinning (shown by arrows in Fig. 7.21) increase with distance from the twin boundary. Normally, slip takes place at a lower stress; hence, twinning is not an important deformation mechanism unless slip is suppressed, i.e., at very low temperatures, in metals with few slip systems (e.g., h.c.p.), or when metals

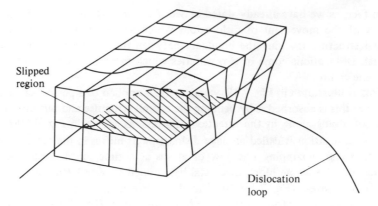

Figure 7.20 Part of a dislocation loop showing edge and screw components (for clarity only part of the crystal is shown).

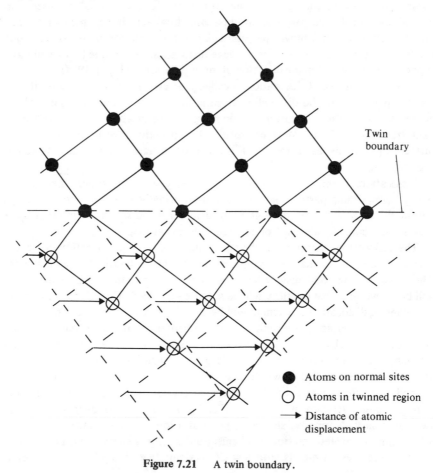

Figure 7.21 A twin boundary.

are subjected to shock loading. (Twinning can take place much more rapidly than slip.)

Twins are frequently seen in the grains of annealed metals, probably as a result of growth from very small twins formed during previous plastic deformation.

7.6   Hardening of metals

The four most important mechanisms for the hardening of metals are hardening due to the presence of grain boundaries, work hardening, solute hardening, and precipitation (second-phase) hardening.

(a) Grain-boundary hardening

The grain boundary acts as a barrier to dislocation movement because there is a sudden change in crystal orientation across the boundary. If the crystal structure is such that there are few slip planes (as in the case of the metals with hexagonal close-packed structures, like zinc), then there is very little chance that slip planes on each side of the barrier will be aligned with each other; thus, the freedom of grains to deform by slip is very limited. The metal may have a high value of yield stress, but will tolerate very little plastic deformation before it fractures.

If there are many slip planes, as in the case of the metals with face-centred cubic and body-centred cubic structures, then there is a fair chance that slip of one grain can be accommodated by slip in a neighbouring grain. A large amount of plastic flow is possible and the metal is ductile. The dislocations still have to overcome the crystal strain field in the region of the grain boundary to pass from one grain to another, and this takes energy. A fine-grained metal has greater *yield strength* (σ_Y)† than a coarse-grained metal. For many metals, including many steels, the yield strength is related to the grain diameter d by the relation

$$\sigma_Y = \sigma_0 + kd^{-1/2},$$

where σ_0 and k are constants.

(b) Work hardening

If a metal is stressed beyond the yield point, then unloaded, and loaded again, the yield point for the second loading occurs at a higher stress than the first time. This effect is progressive, the yield point occurring at a higher and higher stress for each successive cycle of loading, unloading, and reloading. One obtains a stress–strain relationship of the type shown in Fig. 7.22. This repetitive cycle, each time stressing beyond the yield point, occurs when a metal is worked by processes such as rolling, drawing, and forging. The increase in yield stress is called work hardening.

---

†For simplicity, one can assume that the yield strength is similar to the elastic limit.

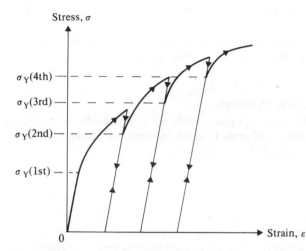

Figure 7.22 Progressive increase of yield stress (σ_Y) with successive loadings – work hardening.

The reason for this is that a very large number of dislocations are generated each time the metal is plastically deformed. When the dislocation density becomes very high, the dislocations interfere with one another and so dislocation movement is hindered. The interference between dislocations can be categorized into three types as illustrated in Fig. 7.23.

1. Dislocations on intersecting slip planes can interact via the release of some of the strain from the disorder around each dislocation to create a dislocation that cannot move in either slip plane (Fig. 7.23(*a*)). This is called a sessile dislocation. It causes a blockage on both slip planes, creating a 'traffic jam' of other dislocations on these planes. The dislocations that are held up will repel each other due to the crystal strain; thus, the 'traffic jam' can rapidly extend for long distances along the slip plane.†

2. The movement of a dislocation passing on a slip plane parallel to a sessile dislocation, or one held up in a 'traffic jam', will be badly hindered by the overlap of the strained region of crystal around the two dislocations (Fig. 7.23(*b*)). This can be described as the effect of the repulsive force between like dislocations.

3. The various slip planes in a crystal run through one another; this means that, when the dislocation density is high, a dislocation will cut through many

---

†This only occurs for dislocations of the same 'sign'. There is actually an attractive force between dislocations of opposite sign, i.e., two edge dislocations, one with the extra plane above the slip plane and one with the extra plane below the slip plane. These will combine and disappear.

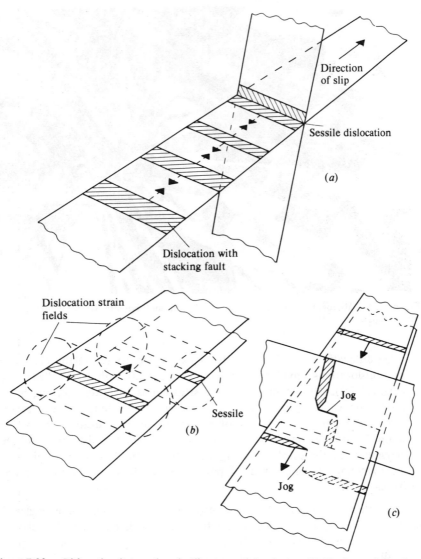

Figure 7.23   Dislocation interactions leading to work hardening. (*a*) Blockage of slip plane by a sessile dislocation (or dislocation tangle); (*b*) overlap of strain fields on neighbouring slip planes; (*c*) formation of jogs when dislocations run through each other.

Figure 7.24 Micrograph of dislocation tangles in molybdenite (transmission electron microscope).

other dislocations that intersect its own slip plane. Each time this occurs, the two dislocations involved both acquire a 'kink', this is called a *jog* (Fig. 7.23(c)). The geometry of the situation is such that the jog in one dislocation is like a small part of the other dislocation. This can cause a serious reduction in the ease of movement of each dislocation, because normally the jog cannot move in the slip plane of the dislocation that has acquired it by the usual unzipping procedure. Atoms have to be displaced by large amounts, leaving a trail of point defects behind the jog like the wake behind a ship.

These three types of interaction eventually lead to large immobile tangles and networks of dislocations that remain when the stress is removed. More havoc is done to the crystal structure with each cycle of loading beyond the yield point, until all ductility is lost and brittle fracture occurs. A micrograph of dislocation tangles in molybdenite is shown in Fig. 7.24.

This type of hardening, in common with grain-boundary hardening, can be applied to almost any metallic system and works in addition to other forms of hardening such as solute hardening.

(c) Solute hardening

Solute atoms, both substitutional and interstitial, will increase the yield strength of a metal. The reason for this is that the misfit of the foreign atom in the crystal creates distortion of the structure. When a dislocation passes through the strain field, the atomic rearrangements associated with dislocation movement cannot take place so easily and movement is hindered.† This is rather similar to the interaction between dislocation strain fields (Fig. 7.23(b)).

The magnitude of the strain field is determined by the degree of misfit of the solute atoms. It has been found that, in many cases, the increase in yield stress is directly proportional to a misfit parameter (Hume–Rothery) based on the comparison of the atomic size, crystal structure, and valency of solute and host species.

We will discuss two examples of solute hardening in the last chapter: namely, substitutional solution Cu–Ni alloys and carbon in martensitic steel (interstitial solution).

(d) Precipitation hardening

This applies to alloys that form a two-phase structure. The presence of a second phase which is of a different crystal structure to the host results in a higher yield stress than a single-phase alloy of the same composition. This is because the dislocations in the host find it difficult, if not impossible, to pass through regions of second phase.

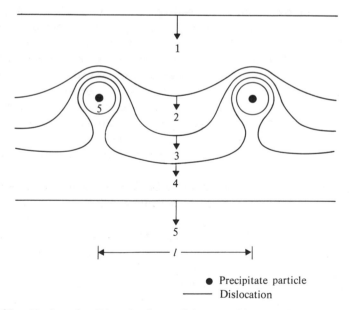

● Precipitate particle
—— Dislocation

Figure 7.25 Pinning of a dislocation by precipitate particles. Numbers indicate the time sequence.

†The defect may actually remove some of the strain existing at the dislocation which therefore becomes trapped at the defect in a potential well.

This effect is called *precipitation hardening* because the alloy is often quenched from a single-phase field for working to form the article; it is then heated to a temperature at which there is enough diffusion for the second phase to form as a finely divided precipitate. This method is used for Al–Cu alloys used in the aerospace industry; the details will be discussed in the last chapter. The time that the article is held at the precipitation temperature is called the ageing time, which leads to the alternative name for this method: *age hardening*.

The interaction between a dislocation and a fairly large precipitate is illustrated in Fig. 7.25. The dislocation is pinned by the precipitates until there is enough stress for it to bow out completely around the precipitate and meet on the other side. The dislocation is then free to continue, leaving a dislocation loop around the precipitate.

There is an optimum heat treatment time for precipitation hardening. If the heat treatment is too short, the precipitates will be too small and will not effectively hinder dislocation movement. If the heat treatment is too long, precipitates will grow larger and there will be fewer of them. The separation between precipitates (l in Fig. 7.25) becomes large and the hinderance to dislocation movement is reduced. The alloy has been overaged.

7.7   Softening of metals — annealing

Work hardening places a limit to the amount of plastic deformation a metal can stand before fracture occurs. However, in many metal working processes, such as the production of thin sheet or wire from large ingots, very large reductions in area are required. Even if the total reduction in area is within the fracture limit, the final part will be hard, with low ductility. This may be what is required if we are making springs and knife blades, but considerable ductility is required of sheet steel that is to be further worked by pressing into shapes for car bodies.

One may use a different alloy for the different uses, but sheet steel would still shatter at the first touch of the press if the work-hardened structure were not altered.

Fortunately this can be easily done by heating the metal to a temperature of between one third and one half of its melting point (in K).† This is called annealing. The first thing that happens at this elevated temperature is that the thermal vibrations assist the dislocations over barriers. In addition, the higher concentration of vacancies aids the removal of jogs and climb of dislocations out of the slip plane to take up a lower energy configuration.

The net result is relief of stress in the metal; this process is called recovery. Yield stress and ductility are not greatly changed at this stage because the number of dislocations is still so large that there is little regular crystal structure and slip is still very difficult. However, recovery alone is useful in increasing the

† This assumes that no precipitates of a second phase form at this temperature.

resistance of the metal to corrosion, because corrosion is accelerated by internal stress in metals.

If the metal is left for a longer time at the annealing temperature, the high diffusion rate due to the presence of a large number of vacancies means that atoms can migrate and take up the lowest energy configuration, which, as we know, is a regular crystal structure.

Small remaining areas of perfect crystal will act as nuclei for the growth of new crystals in the solid state and, as the crystals grow, they will sweep away the dislocation tangles, creating a new polycrystalline grain structure with little memory of the work hardening. This is called *recrystallization*. The greater the damage to the crystal structure by working, the higher will be the solid-state nucleation density. Thus, annealing after a large reduction in area results in a structure that is fine grained.

Recrystallization results in a return to values of lower yield stress and higher ductility comparable with those of the metal prior to working. However, it also implies a greatly refined structure due to the break-up of the coarse cast structure by the cold work.

If the metal is held at the annealing temperature after recrystallization, some grains will continue to grow at the expense of others of less favourable orientation. This results in an increase in the average grain size with time and, also, a gradual drop in yield stress due to the reduction in grain-boundary hardening (Sec. 7.6(a)).

The materials engineer has two routes by which he can achieve the desired balance between strength and ductility of a finished part: he can work up to it by a carefully calculated amount of cold work after the last anneal; or he can anneal down to the final properties, using a part of the final dimensions, by precise control of the final anneal. The former method is the one more often used, as it is more easily controlled and the surface finish is usually better because annealing tends to leave a matt finish due to thermal etching.

The necessity for separate annealing stages can be avoided by working the metal at a temperature above the recrystallization temperature so that the metal does not work harden. This is called *hot working* and is the only way that some very brittle metals with few slip systems can be worked.

7.8 Failure of metals — fracture, creep, and fatigue

To end this chapter on metals, we include a brief comment on how metals fail. Knowledge of the alloy failure characteristics is essential in order to predict the service life of any part subject to mechanical stress.

A part subject to stress above the ultimate tensile stress will fracture. This is the bang that signals the end of a tensile test by the breaking apart of the test piece.

A very ductile metal will show a sharp reduction in area due to a large amount of plastic flow as fracture approaches and a 'neck' forms. In a perfectly

202

plastic material, where no work hardening occurs, this necking continues until the area of the neck is zero and the part has failed. This is ductile fracture.

For a brittle metal, where no plastic flow can occur, there will be no necking. The metal will fail suddenly. In a perfect (non-plastic) material, this would occur at a stress equivalent to the theoretical strength of the solid, the interatomic

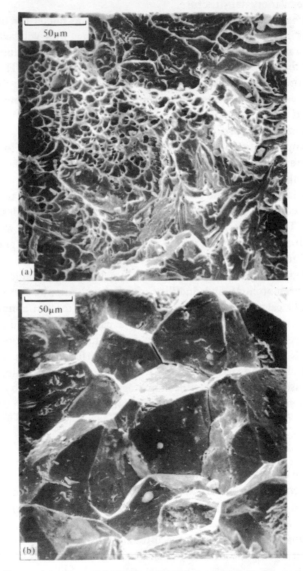

Figure 7.26   Fracture surfaces. (*a*) Ductile 'dimple' fracture in mild steel (initiated by particles of a second phase); (*b*) brittle intergranular fracture in steel. (Scanning electron microscope.)

bond strength. In fact, brittle fracture occurs at stresses two orders of magnitude less than this due to the presence of stress concentrations at radii, notches, or minute cracks in the surface.

The stress applied to the part is amplified many times at the tip of a crack and, when the stress reaches a value equivalent to the strength of the interatomic bonding, the crack propagates rapidly across the solid.

Cracks do not act as stress raisers in ductile metals because the stress is relieved by plastic deformation (creation of dislocations) at the tip.

Most metals in service fall between these two extremes, some necking will occur before the onset of rapid work hardening and brittle fracture.

If stress is applied to a metal at a temperature where diffusion of atoms is appreciable, i.e., in the region of the recrystallization temperature, the internal structure will constantly rearrange itself to reduce the internal stress. This results in a slow progressive deformation of the solid, which extends in the direction of stress. This is called *creep* and can eventually lead to failure. For example, aluminium cannot be used for stressed parts which reach temperatures above 350 °C in car engines because of susceptibility to creep. As a rule of thumb, the more ductile a metal is, the more it will creep. Some plastics will creep at room temperature.

Engineers always design parts so that they are never stressed at anything near to the ultimate tensile strength; in fact, the yield point is not usually exceeded in service, because any plastic deformation would lead to a permanent change of shape. Parts still fail by fracture if they are subject to repetitive cycling of stress, even though the stress is low. This is called *fatigue fracture*. Like brittle fracture, this results from small stress-raisers like cracks and notches in the surface. Although we said that these are blunted by dislocations if the metal is ductile, it appears that they can grow slightly on each cycle (perhaps by work hardening the region at the tip of the crack) until the cross-sectional area of the metal is too small to take the applied stress and normal fracture occurs. The number of cycles to failure increases as the applied stress is reduced, but for many steels there is a value of applied stress, called the fatigue strength, below which no fatigue fracture will occur. It appears that this is not the case for aluminium alloys such as those used for aircraft structures. In this case, there is no safe fatigue strength and the tedious process of testing for a larger number of cycles than the part would experience in service has to be undertaken; the test has to be very carefully designed so that it can accurately simulate stresses encountered in service. The necessity for this type of testing is indicative of the failure of materials scientists to understand fully the causes of fatigue fracture.

Micrographs of different fracture surfaces are shown in Fig. 7.26.

8. Non-metals

8.1 Introduction to mechanical properties

We will now consider non-metallic solids. The treatment will be rather different from that of metals because the movement of dislocations under applied stress, which is crucial to the mechanical properties of metals, is of little or no importance in determining the mechanical properties of non-metals.

By definition, the atomic bonding in non-metals may be covalent, ionic, van der Waals, or any combination of these three. The ceramics and glasses that we will consider tend to have three-dimensional covalently-bonded structures in which dislocation glide is impossible because of the strength and directional nature of the covalent bond. If plastic flow is impossible, then the material cannot usually be worked. The final part has to be made from the ingredients in one process. For this reason, the manufacturing process is intimately linked with the properties of the material and so will be discussed in the appropriate part of this chapter. Just a few examples are saved for the last chapter on materials engineering.

Glasses and high polymers (plastics) can be worked at elevated temperatures, but plastic flow is not due to dislocations. For glasses, plasticity arises from the freedom of bonds to rotate at high temperatures, giving liquid-like flow. High polymers that can flow are composed of long hydrocarbon chains which are intertwined together and held by van der Waals bonding. Flow occurs by the slipping of chains with respect to each other.

This chapter is divided into three sections: ceramics, glasses, and high polymers, although the distinction between ceramics and glasses is rather blurred. Some ceramics, porcelain, for example, are largely composed of glass, and some modern processes have been developed to create ceramics from glass (Pyroceram).

8.2 Ceramics

Because of the complete absence of plastic flow, ceramics have no ductility, although they often have a high modulus of elasticity (high stiffness) and fracture can occur at very high values of stress.

If the ceramic is glassy, then fracture is brittle fracture, as described in Sec. 7.8. This is dominated by stress raisers such as cracks and notches in the surface. At a critical value of stress, these propagate rapidly across the surface and solid fails. This also applies to the pure glasses.

If the ceramic is a single crystal, then fracture tends to occur along the close-packed planes. This is called *cleavage* and occurs because the variation of properties with orientation in a single crystal controls crack propagation and confines it to the most widely separated (low index) planes. The most well-known example of cleavage is the cutting of gemstones — especially diamonds — to many beautiful variations on the crystalline shape (called the crystal *habit*).

Cleavage occurs for layered sheet structure ceramics such as mica and graphite (Secs. 2.14(*a*) and 2.14(*b*)) along the planes of weakness, where van der Waals bonding holds the covalently-bonded sheets together.

If the force on a ceramic or glass is compressive rather than tensile, cracks and flaws tend to close up and many stress raisers are inoperative. This has the result that these materials are much stronger in compression than in tension; careful design (especially in the field of civil engineering) makes use of this fact, as we will see later.

We will now discuss the properties of some ceramics arranged into groups by method of manufacture. The groups are pottery, refractories, abrasives, special ceramics, and cement.

(a) Pottery

Pottery is made from a mixture of clay (kaolinite) and other minerals such as sand and felspar containing varying amounts of silica (SiO_2) and alumina (Al_2O_3) with other metal oxides. As discussed in Sec. 2.14(*b*), this mixture is made plastic by the addition of water, which acts as a lubricant, allowing the covalently-bonded silicate sheets to slip over one another. Items are made either by wet plastic forming (throwing or turning pots), which is mainly done by a machine nowadays, or by slip casting, where so much water is added that a slurry is formed that can be poured into moulds (Fig. 8.1). The moulds are made from plaster of Paris, which soaks up some of the water, leaving a layer of clay on the inside when the surplus is poured out.

Excess water is removed after either process by allowing time for evaporation; the piece is then fired at a temperature at which more water and other volatiles are driven off, and the silica fuses to form a glassy structure.

A glaze (which is a suspension of high silica content) can then be applied. After a second firing, this forms a glassy surface which closes up the pores and cracks in the surface. This can increase the mechanical strength, especially if the glaze has a lower rate of thermal expansion than the pot. In this case, when the pot cools it shrinks more than the glaze wants to; this creates a compressive force in the glaze. This increase in strength gives a resistance to spalling. Spalling

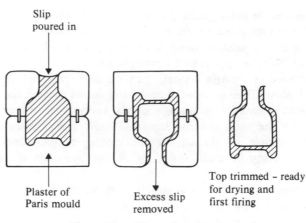

Slip
poured in

Plaster of
Paris mould

Excess slip
removed

Top trimmed – ready
for drying and
first firing

Figure 8.1   Slip casting of pottery.

is the break-off of areas of surface due to forces arising from thermal expansion
of the surface layers of a material of very low thermal conductivity.

Pots can be divided by the firing temperature into the following groups.

1. *Earthenware*   Low firing temperature, up to 20 per cent porocity, e.g.,
   brick, tile, and tubes (drains), made from cheap natural deposits.
2. *Stoneware*   Medium firing temperature, less than five per cent porocity,
   pleasant appearance, and acid resistance with Ba compounds for chemical
   plant.
3. *China and porcelain*   High firing temperature, very dense, and high glass
   content (translucent). Bone china: 45 per cent bone ash, high strength.
   Porcelain: less SiO_2, more dense, lower strength. Porcelains of special
   composition are used as electrical insulators (more of this in Chap. 9).

(b) Refractories

A major user of refractory oxides is the steel-making industry, where they are
used for lining furnaces and ladles. The porous structure of insulator bricks
enables very low thermal conductivities to be achieved (an order of magnitude
less than fused silica). The bricks on the inside surface that are in contact with
molten steel at over 1600 °C have to be resistant to thermal shock and must not
be chemically attacked by the slags on the surface of the steel.

The binary phase diagrams† of two refractory systems, the $SiO_2-Al_2O_3$ and
the $MgO-Al_2O_3$ systems, are shown in Fig. 8.2.

In both cases, an intermediate compound is formed: mullite ($3Al_2O_3.2SiO_2$)
in one case and spinel ($MgAl_2O_4$) in the other. We have already met the spinel
structure when discussing the ceramic magnet magnetite (Sec. 4.15).

---

†These are actually pseudo binary phase diagrams because they are not between elements,
but between compounds.

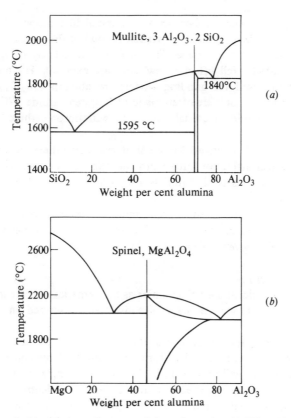

Figure 8.2 Pseudo-binary phase diagrams of the refractories. (a) SiO_2–Al_2O_3, (b) MgO–Al_2O_3.

The refractories from the silica–alumina system are rather more resistant to thermal shock and easier to make, but they must be on the alumina side of the mullite composition or they would not be able to withstand the temperatures of the steel-making process. Therefore, the refractories used are a mixture of mullite and alumina. Refractories of the magnesia–alumina system have a greater resistance to slags, spinel has a greater thermal shock resistance than pure MgO.

(c) Abrasives

These are used for grinding wheels and metal cutting tools. They are also used in the form of a powder mixed with lubricant as lapping, grinding, and polishing pastes. The quality of the finish depends on particle size.

For grinding and cutting applications, sharp fragments of a very hard material (strong three-dimensional covalent bonds) are held in a soft matrix which may be ceramic, glass, plastic, rubber, or a soft metal like copper. The matrix is quickly worn away, leaving the hard particles proud of the surface so that a film of lubricant can cover the surface. The surface is constantly being renewed as used abrasive particles fall off and new ones are exposed. The most common abrasives that are used for grinding wheels are alumina (Al_2O_3) and silicon carbide (SiC). Cutting tools are often made of tungsten carbide (WC) in a matrix of chromium. The hardest material is diamond with boron carbide (B_4C) a close second.

Covalent bonding is strongest for small atoms where there are no inner shells to mask the nuclear attractive force from the valence electrons. This is why the carbides predominate as hard materials with 'carbon carbide', diamond, as the ultimate in hardness.

By their very nature, hard solids cannot be melted easily, so parts are normally fabricated by pressing of powders followed by sintering. Details of the methods used will be given in Chap. 9.

(d) Special ceramics

Little needs to be said of special ceramics as the ceramics that are important for particular physical properties have already been described in the relevant sections of this book.

One can divide special ceramics into materials that are used in the polycrystalline form and can, therefore, be fabricated by conventional techniques and those that will only work in the form of a single crystal. Ceramics of the first type include ferroelectrics for capacitor dielectrics, ferrites, and phosphors. Single-crystal ceramics include semiconductors for electronic devices, lasers, photoconductors, piezoelectric materials and alkali halides for infra-red prisms and lenses.

(e) Cement

This is a case in which the ceramic is made by an *in-situ* chemical reaction. In this case, the reaction is between powdered minerals and water. The reaction is exothermic (heat is given out), which can create problems for very large structures.

This is one of the cheapest structural materials available, because it uses some of the most commonly occurring materials on the earth's crust: limestone (CaO) and clays (containing SiO_2 and Al_2O_3). The clay and limestone are ground to a powder, fired to drive-off the water of crystallization, and then reground to form a very fine powder.

The reaction is between water and the silicates in the mixture, for example, calcium silicate ($3CaO.SiO_2$). When the powder is mixed with water to form a slurry, some of the silicates dissolve; then, over a period of time, new crystals grow, nucleated on the solid particles in the slurry. The new crystals are layered-structure hydrated silicates in the form of long fibres. The water is

Figure 8.3 Crystallized cement paste showing the fibrous nature of many of the crystals (scanning electron microscope).

consumed by incorporation into the crystal structure (where it is needed to fill positions vacated during the original firing process), to satisfy bonding between the silicate sheets. Cement does not dry out. When the fibrous crystals nucleated on neighbouring grains interlock, they form a solid three-dimensional structure and the cement has reached its maximum strength. A micrograph of cement particles is shown in Fig. 8.3.

Portland cement, which is mainly SiO_2 and CaO with a little Al_2O_3 and FeO, takes a month to set.

High-alumina cement, which is mainly Al_2O_3 and CaO, takes only one day to set. This is very attractive for the building trade, but an unstable silicate is formed in high-alumina cement which decomposes slowly to a more stable form with a less interlocked structure. This results in a loss of strength, but this is not serious if the water content of the slurry is carefully controlled. However, if there is too much water, the loss of strength may be sudden and catastrophic.

Figure 8.4 A model of the structure of a tetrahedrally bonded glass.

Cement and concrete (gravel, sand, and cement) in common with many ceramics are much stronger in compression than in tension. This must be allowed for in design, but compression can be 'built in' by casting around tensioned steel rods (pre-stressed concrete).

8.3 Glasses

The basis of most glasses is silica (SiO_2), although there are other glass 'formers' such as the oxides B_2O_3, GeO_2, P_2O_5, As_2O_5, and As_2O_3, but these are usually used as additives to silica.

In these oxides, diffusion is so sluggish that, if they are cooled quickly enough from the liquid state, no crystalline structure is nucleated. A three-dimensional random network is formed in which, in the case of silica, each silicon atom is at the centre of a tetrahedron of four oxygen atoms and each of these oxygen atoms is joined to another silicon atom. All bonds are satisfied, just as they are in the crystalline forms of silica such as quartz, but one cannot predict from the position of one silicon atom exactly where the neighbouring silicon atoms will be. This uncertainty in bonding angle is similar to the situation existing in liquids. In fact, one can quote a viscosity for glasses, arising from the freedom of atoms to take up a range of positions with respect to one another. The viscosity varies greatly with temperature. To be workable, the glass has to have a viscosity of 10^3 N s m^{-2}† or less. The softening point is the temperature at which the viscosity is 10^7 N s m^{-2}. At temperatures below this, rotation of the bonds is difficult, but internal stress in the glass can be relieved (annealing) down to a viscosity of 10^{12} N s m^{-2}. A model of the structure of a tetrahedrally bonded glass is shown in Fig. 8.4.

Pure fused silica has a very high working temperature and so fabrication of parts is both difficult and expensive. This limits its use to applications where its properties of high resistance to thermal shock and chemical attack, low thermal expansion, and low electrical conductivity (no metal ions) are of prime importance. Normally, modifiers such as the alkali metals are used. These have only one valence electron, so can only bond to one atom in the structure. This breaks up the three-dimensional network of interconnected bonds, giving a more open structure with a much lower softening temperature. Soda-lime glass (it contains CaO and Na_2O) is easy to work and makes up the vast majority of the glass used, but it has poor resistance to thermal shock and chemical attack. An alumina—borosilicate glass (contains Al_2O_3, B_2O_3, and a little Na_2O) is intermediate in properties and cost between soda-lime glass and silica, this has the trade name of Pyrex.

Other oxides may be added to the glass to alter its properties. For example, lead oxide (PbO) is added to increase the refractive index for decorative cut glass to give the glass more sparkle.

---

†The force needed to move one layer past another at a fixed velocity.

An almost pure silica glass, known as Vycor, can be made reasonably cheaply using a workable two-phase glass (containing B_2O_3 and Na_2O) to make the required shape. Then the chemically active phase containing the sodium and the boron is leached out by chemical attack, leaving a honeycomb of silica which can be subsequently sintered to remove the porocity with a 30 per cent reduction in volume.

Glass ceramics (called Pyroceram) have been made in special glasses containing crystals of TiO_2. These can be worked to shape and then heat-treated at a high temperature, at which fine crystalline precipitates nucleate around the TiO_2 crystals. The resulting ceramic has a low coefficient of expansion and high strength, because the precipitates act as crack stoppers.

The number of working processes for glasses is large and includes many similar to metal-working processes such as casting (normal and centrifugal – e.g., the back ends of TV tubes), rolling, drawing, and pressing.

Jam jars and other bottles are made by the press and blow method, as illustrated in Fig. 8.5.

Plate glass, which originally had to be ground and polished to achieve the required surface finish, is now made by the Pilkington process (Fig. 8.6), in which the glass flows on, and replicates, the mirror-smooth surface of molten tin. The mirror-smooth glass sheet is drawn off by very smooth rollers. In this process, in common with many others, the glass has to be cooled (annealed) very slowly so that internal stress arising from differential shrinkage due to the material's poor thermal conductivity can be relieved.

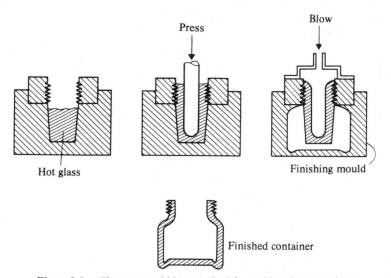

Figure 8.5   The press and blow method for making glass containers.

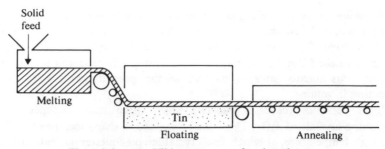

Figure 8.6   The Pilkington process for plate glass.

Internal stress is deliberately built into glass for car windows† by rapidly cooling the surface with air jets while the core is still plastic. On cooling, the core contracts more, placing the surface in compression. In an accident, the stress is released and the windows shatter into very small fragments which cannot act as damaging projectiles.

Glass fibres are made by feeding glass in the form of marbles into a platinum tank that is heated by the passing of a very high electric current through the tank. Small jets are placed at the bottom of the tank and the glass is drawn out from these on drums rotating at very high speed.

8.4   Composite materials

Glass fibre is often a component of composite materials in which the different properties of materials are combined to give a material that performs better in service than either of the two components. Perhaps the earliest example of this is the combination of clay and straw to make bricks.

Glass fibre reinforced plastic (GRP) is the combination of strong but brittle glass fibres in a matrix of ductile but weak plastic. Stress is transferred by deformation of the plastic to the glass fibres if the bond between fibre and matrix is stronger than the plastic. A good bond is best achieved by a high-temperature heat treatment.

The best composite materials have continuous fibres aligned with the direction of stress. The modulus of elasticity of the composite is the sum of the moduli times the concentrations of the two components; the ratio of the loads carried by the components is the ratio of the moduli times concentration. As glass can have a modulus three orders of magnitude greater than plastic, this means that the elastic modulus of GRP is just the modulus of the glass times its concentration, whilst 99.9 per cent of the load will be carried by the glass fibres.

---

†Laminated glass is often used now for windscreens so that when it shatters visibility is retained. The large fragments are held by the plastic 'meat' in the sandwich.

Figure 8.7 A carbon-fibre–epoxy-resin composite fractured by bending. The fibres have been pulled out by tensile strain (region on the right) but not by compressive strain (region on the left).

The ultimate strength of composite materials is high because cracks cannot propagate from the brittle to the ductile components. A micrograph of a fracture surface of a composite is shown in Fig. 8.7.

Recently, glass fibres have been replaced by carbon fibres, which have a continuous chain structure that gives them enormous strength (unlike the sheet-structure form of carbon: graphite).

8.5 The high polymers (plastics)

The reader will not need to be reminded of the large use made of plastics in fabrication of domestic articles. The great advantages of plastics are low cost, ease of production of very complex shapes, very accurate control of dimensions, good surface finish, flexibility, ease of control of colour and/or transparency, and low electrical and thermal conductivity. The disadvantages are low strength and drastic reduction in strength (or degradation) at elevated temperatures.

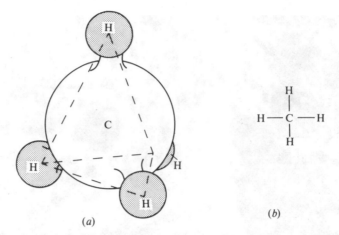

Figure 8.8 Methane, the building block of plastics. (*a*) Tetrahedral structure; (*b*) symbolic representation.

The plastic structure is based on the chemistry of carbon combined mainly with hydrogen and sometimes nitrogen, oxygen, and the halogens. The properties of plastics depend greatly on the precise shape of very large molecules. The various elements are only important because they affect the bonding that determines the shape of the molecule. The simplest hydrocarbon is methane, which is carbon tetrahedrally bonded to hydrogen. This is similar to the silicon tetrahedron and an idea of the shape of the molecule is shown in Fig. 8.8, along with the normal shorthand way of representing the structure by chemical symbols. We will use the latter notation, but it must always be remembered that the bonds are really tetrahedral.

The secret of polymerization to form plastics which have very long chain molecules is that a carbon atom can be bonded to another carbon atom by one, two, and, occasionally, three bonds (each bond has two electrons).

If there is a double bond, then, if this can be converted into a single bond, it releases two bonds for bonding to neighbouring molecules and *polymerization* can begin. For example, polymerization of the simplest of the building blocks of the polymers (which are called *mers*), ethylene, is shown below:

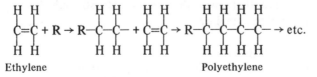

The radical R is called an initiator, as it opens up the original double bond by forming a bond with the carbon atom which leaves one bond free to bond with the next molecule. Polymerization will continue until the chain meets the active end of another chain or another radical which acts as a terminator. The polymer chains are held together by van der Waals bonding and mechanical intertwining, which is considerable as the chain is free to rotate about any of the bonds. The strength increases with chain length, so long chains are desirable and molecules with more than one thousand carbon atoms are common.

Some of the most popular plastics, which, together with polyethylene, make up 65 per cent of all plastic sold, are based on the mers listed below:

```
H H H
| | |
C=C-C-H     propylene → polypropylene,
| | |
H H H


   H Cl
   | |
H-C=C-H      vinylchloride → polyvinylchloride,


H H
| |
C=C         styrene → polystyrene.
|
   —(the benzene ring)
```

Rubber is also a high polymer based on the mer isoprene:

```
H H  CH3 H
| |   |  |
C=C——C==C     isoprene → rubber (polyisoprene).
| |      |
H H      H
```

This is called an elastomer as it is rubbery at room temperature.

As they are cooled, the chain-structure high polymers can go through three types of mechanical deformation:

1. *Viscous or plastic*   The chains slide past one another to give very large plastic deformation.
2. *Rubbery*   The chains uncoil when stress is applied but there is not enough thermal activation for them to slide past each other. Deformation is elastic. This applies particularly to polymers with bulky side groups which tend to interlock.
3. *Glassy*   The molecules are locked in place, resulting in the properties of high elastic modulus but no ductility.

The mode of mechanical deformation is sensitive to the strain rate as well as temperature because the molecules take time to unwind. Thus, a very rapid application of load can result in brittle fracture of a polymer which would be in the rubbery range at low rates of strain.

The ultimate tensile strength of the chain polymers of the type we have just discussed is two orders of magnitude down on the stronger solids (steel and glass), ranging from $20 \, \text{MN m}^{-2}$ for polyethylene to $80 \, \text{MN m}^{-2}$ for poly-

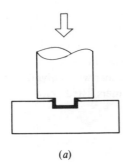

(a)

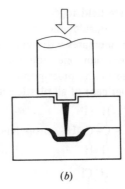

(b)

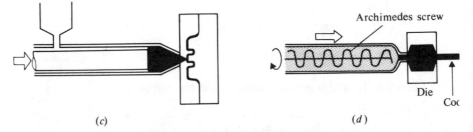

(c)   (d)

Figure 8.9 Processing of plastics. (a) Compression moulding; (b) transfer moulding; (c) injection moulding; (d) extrusion.

amide.† The elongations that can be achieved are very large, up to 800 per cent for low-density polyethylene. (Polystyrene is an exception with less than two per cent elongation.)

The simpler chain molecules, such as polyethylene, can align themselves by wrapping backwards and forwards to form a regular crystal structure over quite large regions of the solid, but the crystalline perfection is not quite as good as can be achieved in inorganic solids.

The chain-structure polymers are called *thermoplastics* and they can be melted, cast, and then remelted.

†Sometimes called 6/6 nylon which is a chain of the unit

$$-\underset{\underset{H}{|}}{N}-\underset{\underset{O}{\|}}{C}-$$

interspaced with short lengths of

$$-\underset{\underset{H}{|}}{\overset{\overset{H}{|}}{C}}-\underset{\underset{H}{|}}{\overset{\overset{H}{|}}{C}}-$$

There is also a second class of plastics, called the *thermosetting plastics*. These are formed *in situ* by a chemical reaction in the mould, which creates a three-dimensional network rather than a chain structure. In this case, organic molecules are used in which there are two double bonds and, therefore, two places where other molecules can be attached. We have already seen one mer of this type: isoprene. In fact, the chain structure of the very soft rubber is partially converted into a three-dimensional structure by using sulphur to create a cross-link between chains via the spare double bond. This process is called vulcanization.

Rubber for car tyres is only partially cross-linked, but the more typical thermosetting plastics can be recognized by higher rigidity and mechanical strength than thermoplastics (by virtue of the three-dimensional network structure). This also means that they retain their rigidity to higher temperatures, but eventually they char and decompose.

Both types of plastic are often mixed with a large percentage of fillers, which further reduce costs but can also increase strength. Typical fillers are naturally occurring materials like wood, cellulose, mica, asbestos, and graphite.

Wood is a honeycombed natural high polymer and the difference in strength of the different types of wood arises from the difference in the geometry and degree of refinement of the cellular structure.

The various methods of moulding plastics are summarized in Fig. 8.9. Thermosetting plastics are moulded by compression of the powdered components into a heated mould, either directly or by transfer from a holding pot. Thermoplastics are melted and either injected by means of a ram into a mould or extruded by forcing the liquid through a small aperture (die), using pressure applied by an Archimedes screw which feeds the plastic to the die.

9. Materials engineering case histories

9.1 Introduction

Now that the ideas behind materials engineering have been introduced, we will use this chapter to give more specific examples of the factors affecting the choice of materials, and the materials engineering necessary to achieve the required properties.

Many applications have already been given in the relevant sections throughout this book, so examples have been chosen that illustrate points that have not been covered in any detail earlier in the text.

Cost is an important factor in materials choice. This factor has not been given much prominance so far, but it must feature here. Another factor that is now becoming important with the dwindling of world resources is the energy used in manufacture and the ease with which the materials can be re-used after service. These factors do not yet greatly influence materials choice and so do not feature here, although they may become very important in the future.

Three examples from the field of mechanical engineering are discussed: materials for motor-car components, materials for aircraft components, and high-temperature materials. The two other examples are materials for electrical conductors and insulators for power transmission lines, and the growth and purification of single crystals.

As the use of materials in the field of civil engineering is, as yet, fairly unsophisticated, we hope that this will have been covered in our discussions earlier in the book and also by discussion here of steel making.

9.2 Case 1. Metals for motor-car components

The requirements for motor-car components are enormously varied, but most important mechanical and structural parts can be made from alloys of just one metal: iron. The advantages of iron alloys are primarily low cost and the tremendous variety of properties, including very high tensile strength, that can be achieved from varying the composition of alloys and the heat treatment. No other alloy system can compete on cost unless the requirements demand a superior resistance to corrosion (unfortunately not normally a requirement for car bodies), a superior strength to weight ratio, or service at high temperatures.

(a) The iron–carbon system

The iron-rich end of the iron–carbon phase diagram is shown in Fig. 9.1. We have mentioned in the section on crystal structures (Sec. 2.7) that iron is b.c.c. up to 910 °C – this is the α phase (called *ferrite*) – and f.c.c. between 910 °C and 1400 °C – this is the γ phase (called *austenite*). The δ phase is again ferrite (b.c.c. structure).† The phase diagram stops at the intermediate compound, iron carbide (Fe_3C).

The carbon forms an interstitial solid solution and, as the gaps in the f.c.c. lattice are much larger than those in the b.c.c. lattice, far more carbon can go into solution in austenite (up to two per cent) than in ferrite (up to 0.03 per cent). Pure iron (either austenite or ferrite) is ductile but with a relatively low tensile strength ($\sim$90 MN mm^{-2}); this is traditionally called 'wrought iron' because of the ease with which it is worked.

Alloys with a high carbon content (2 to 4.5 per cent carbon; white cast iron, malleable iron, grey iron, nodular iron) are used for forming castings of all kinds. Often they contain large amounts of the brittle carbide Fe_3C called *cementite*. If, instead, the carbon is in the form of graphite, this makes a very machineable alloy. Cast iron is strong but often brittle, so that it cannot be easily worked; the final shape is achieved by machining.

The steels have a carbon content of between 0.05 and 2 per cent and retain some of the ductility of wrought iron whilst being up to 20 times stronger. This is achieved by control of the mix of ferrite and cementite.

Some idea of how this is done can be obtained by considering the eutectoid transformation:

$$\gamma \text{ (austenite)} \xrightarrow{\text{cooling}} \alpha \text{ (ferrite)} + Fe_3C \text{ (cementite)}.$$

The steels on the iron-rich side of the eutectoid composition ($<$0.8 per cent carbon – called hypoeutectoid) are by far the most used because of their high ductility. If we cool slowly from the single-phase austenite (γ phase) region (Fig. 9.1(b)), the α phase starts to form at the grain boundaries where diffusion rates are highest. When the temperature falls below the eutectoid temperature, the remaining austenite (γ) transforms to an interleaved or lamellar mixture of the ductile α phase and the brittle Fe_3C. This particular mixture of the two phases is called *pearlite*. The pearlite adds strength because the lamellae of the brittle cementite limit slip in the ferrite regions, but some ductility remains because a considerable proportion of the alloy is still ferrite and the cementite is very finely divided.

---

†There is no β phase. Originally, it was thought that the loss of ferromagnetism at the Curie temperature (770 °C) was due to a new phase and this was given the symbol β.

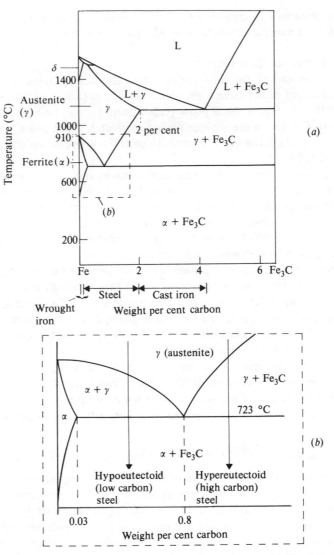

Figure 9.1 The iron-rich end of the iron–carbon phase diagram. (*a*) The full diagram; (*b*) the eutectoid region.

High-carbon steels (>0.8 per cent carbon – called hypereutectoid) have a higher strength, hardness, and wear resistance. In this case, the brittle cementite (Fe_3C) forms at the grain boundaries on cooling from the single-phase γ (austenite) region to the $\gamma + Fe_3C$ region. As in the case of hypoeutectoid steels, the remaining austenite transforms to pearlite on cooling below the eutectoid temperature. The coarse cementite that formed in the $\gamma + Fe_3C$ region gives the alloy greater strength and hardness but at the expense of ductility.

So far, we have just considered slow cooling. If we quench steel rapidly from the austenite region to well below the eutectoid temperature, there is only sufficient time for the smallest amount of carbide to form as formation of this phase requires considerable diffusion. Nevertheless, the iron will transform from the f.c.c. austenite to the b.c.c. ferrite as the change in crystal structure does not require diffusion in order to take place. Now we have a problem. Where does the carbon go? We have up to two per cent carbon to be accommodated in a lattice which can normally only take 0.03 per cent. What happens is that a new non-equilibrium phase is formed with a highly strained b.c.c. crystal structure to accommodate the carbon in an interstitial solid solution. This phase is called *martensite* and is exceptionally hard. It owes this hardness to three cumulative effects.

1. Very small grain size (as we have quenched to well below the entectoid temperature).
2. Up to ten per cent distortion of the b.c.c. lattice by incorporation of the carbon as an interstitial solid solution.
3. Some very small precipitates of the carbide Fe_3C are formed giving precipitation hardening.

Thus, the three main ways of hindering dislocation movement are combined in one metal. Matensitic steel is used where the greatest strength and hardness is required (for example, a knife blade), but sometimes the steel is heated to allow some more carbide to form. This process increases ductility with some loss of hardness and is called *tempering*.

We will now consider how these properties can be used for manufacture of motor car components.

(b) The use of iron alloys

The structural parts of the car engine, such as the crank case, the cylinder block, and the cylinder head, have very complex shapes. These can only be achieved by casting and machining. The problems with casting are that we are liable to have a large grain size and segregation leading to non-uniform properties; in addition, the presence of the carbide phase produces a very brittle structure. If the casting is cooled slowly, graphite forms instead of carbide, leading to much improved properties. Grey iron is used for parts such as those mentioned above and also piston rings, tappets, camshafts, and manifolds. The graphite microstructure is optimized for ease of machining. Another cast iron, called ductile iron, with a graphite microstructure optimized for strength, has such a refined structure that it can be used for mechanically stressed parts such as crankshafts and rocker arms.

The technique used for sand casting is shown in Fig. 9.2. A pattern of the required part is made of wood and is used to make a cavity in a special fine sand.

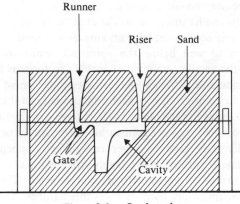

Figure 9.2 Sand casting.

The metal is poured via a runner and the flow rate controlled by a small opening (the gate). Air is expelled through the riser and metal is poured until the riser fills, providing a hot head of metal to contain the shrinkage.

Parts that have to have a high modulus of elasticity and high strength, such as springs, will be made from high-carbon (hypereutectoid) steels. They will be worked down to the final shape by hot rolling (Fig. 9.3) which breaks up the coarse cast structure of the ingot. Bearings and gears that need high wear resistance are also of high-carbon steel, perhaps with a martensitic surface layer or other structures produced by special heat treatments or carburizing.

Parts such as connecting rods and suspension components, subject to cyclic stress, are often forged by hammering in a die whilst hot. This also results in

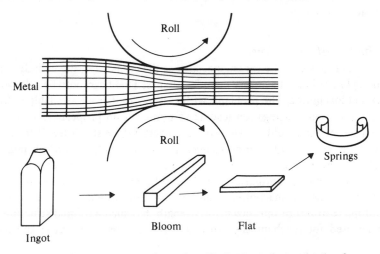

Figure 9.3 Rolling steel. (*a*) Deformation; (*b*) shapes used on reduction for springs.

refinement of the grain structure with the grains aligned favourably to counteract the stress encountered in service. Forging can produce surface free of stress-raisers such as cracks so reducing the chance of failure.

Sheet to make the car body will be a low-carbon steel, as ductility is important; this will also be produced by hot rolling of a large ingot. The sheet is then annealed and pressed into the final shape. A high degree of ductility is required in order to deform the metal sufficiently to achieve the required shapes.

The car body will be fabricated from sections by welding, where the metal to be joined is melted to form an intimate bond. Allowance has to be made in the design for the fact that the weld will not have the strength of the work-hardened steel around it because it has solidified from molten metal and will therefore have a cast structure.

Some parts, such as the exhaust system, really require a corrosion-resistant alloy as the heat and moisture present on the inside are ideal conditions for corrosion. An alloy such as stainless steel, which contains a large percentage of chromium, would be best suited to this application, but normally this is ruled out by the extra cost of the material and the greater difficulty of welding. One needs to weld stainless steel in an oxygen-free atmosphere. This is done by flooding the area around the arc struck between an electrode and the metal to be joined with argon. This is called *argon arc welding*.

Let us now consider alloys other than those based on iron to see in what areas they can compete.

(c) Other alloys

Weight is extremely important in the reciprocating parts of the engine such as the valve gear and the pistons because their inertia opposes the constant reversal in direction of motion. This is reduced by design in the case of the valve gear because usually only steel will give the necessary strength, but the pistons are another matter. They have a complex shape and are therefore machined from castings, but they do not need great strength. Thus, this was the first area where iron was replaced by another metal, aluminium in this case.

Aluminium has a low density but has a strength only one quarter that of iron; it is also five times the cost of iron and has a lower melting point. Use can be made of the last property because aluminium parts can be cast in steel moulds by injection of the molten metal into the mould under pressure. This is called die casting, and the castings produced can be much finer and of more intricate shape than for a sand casting. As a result, less machining is required to produce the finished part. Normally, the die castings are made from two-phase aluminium alloys containing 10 to 20 per cent in total of metals such as Si, Cu, Fe, and Mg. The alloys have a much greater strength than pure aluminium due to solution or precipitation hardening.

Aluminium alloys such as Al–Si are now replacing grey iron for engine components because of the low weight, higher thermal conductivity (so that the engine runs cooler), and refinement of casting techniques. This is especially so for applications where saving of weight is important, such as in motorcycle and aircraft engines. We have already studied the Al–Si phase diagram (Fig. 7.11) and noted that a two-phase structure is formed, with the phases distributed as a fine eutectic mixture plus larger crystals of one or the other of the components, depending on which side of the eutectic composition the alloy is. The two-phase structure gives the alloy a high yield strength (compared with aluminium) and grain refinement is sometimes achieved by controlling nucleation of the solid. This is done by the addition of phosphorus, which reacts to form crystals of aluminium phosphide that act as nuclei in the melt.

Single-phase aluminium alloys with less than two per cent of additions such as Mn, Mg, and Cr to give substitutional solution hardening are used in the form of sheet for bus and lorry bodies, where work hardening, as a result of the rolling and bending, adds to the solution hardening to give the required yield strength. These alloys score over steel in weight, corrosion resistance, and ease of working due to a much greater ductility. This latter property means that, for example, the sides and bottom of a beer can can be pressed from a single sheet of aluminium, whereas a steel beer can has to be made of tube with the bottom soldered on. This is an example where the higher cost of the raw material can often be offset by the lower cost of production.

Another case where this holds is in the field of bearings. Steel bearings (ball, roller, needle, etc.) are complex fabrications that have to be made to very exacting tolerances, but for most applications these can be replaced by journal bearings. The bearing surface is formed from a soft metal such as alloys of copper and lead. Copper and lead are completely immiscible in the solid state and so the lead separates into islands in a pure copper matrix. In service, the lead from islands that are exposed on the surface is smeared by the rotation of the parts over all the surface of the bearing to form a layer with a very low coefficient of friction. The copper acts as a mechanically rigid support. As the bearing wears, new lead islands are constantly being exposed, so that the layer of lead is maintained.

9.3 Case 2. Metals for aircraft components

The first aircraft was made from wood and fabric with some metal fittings. This gave the necessary area to generate lift whilst keeping the weight low. When larger planes were planned, the stresses involved became too great for this type of construction: a structure with a metal frame and cladding was needed. An aeroplane made of steel would not get off the ground and one made of aluminium would not be any better because of the low strength of pure aluminium.

This problem was solved by using an aluminium–copper alloy that has a strength to weight ratio much greater than steel. This alloy owes its strength to precipitation hardening and has the trade name of Dural.

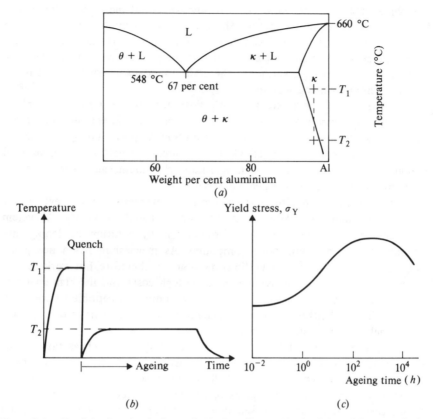

Figure 9.4 Precipitation hardening of Al–Cu alloys. (*a*) Aluminium-rich end of the Al–Cu phase diagram; (*b*) temperature–time relationship for the heat treatment; (*c*) change of yield stress with ageing time.

We can see how this is achieved by looking at the aluminium-rich end of the Cu–Al phase diagram (Fig. 9.4(*a*)). The diagram is complicated at the copper-rich end, but we are not concerned with this, so we can regard the aluminium-rich end as a eutectic between the intermetallic compound $CuAl_2$ (θ phase) and aluminium with a little copper in solid solution (κ phase).

An alloy with about 4 per cent copper which had been worked and annealed with relatively slow cooling would have a phase structure comprising grains of the κ phase with a coarse distribution of $CuAl_2$ at the grain boundaries. In this form, the alloy is ductile and can be worked easily to the required shape. The alloy is hardened by following the heat treatment indicated in Fig. 9.4(*b*). The part is heated to a temperature (T_1) in the κ field and time is allowed for all the θ phase to disappear as the copper goes into solid solution. Then a rapid quench to temperatures where diffusion is too slow for the θ phase to form freezes the alloy in the single-phase form. The temperature is then raised to T_2 in the two-phase field, where diffusion is relatively rapid, and the θ phase forms as a finely divided coherent precipitate. The word coherent implies that the crystal lattice of the precipitates is closely linked to the matrix crystal structure. This gives a much greater hardening effect than if the precipitate formed a separate grain, effectively leaving a surface at the grain boundary where dislocations can run out rather than being held up.

The increase of yield strength with time due to the growth of θ-phase precipitates is shown in Fig. 9.4(c), where one can see that for very long times one can start to lose strength as the distribution of precipitates becomes too coarse. The aluminium–copper alloys used for aircraft structures may have small amounts of silicon, manganese, magnesium, chromium, and zinc to tailor properties to the exact requirements in service.

Magnesium alloys are also used for aircraft as magnesium has an even lower density than aluminium. Alloys that are used are based on the Mg–Al system with less than ten per cent Al. Hardening is by solution hardening and precipitation of an intermetallic compound. As magnesium has a hexagonal close-packed structure, it has few slip systems and is, therefore, hot worked.

The use of magnesium alloys is limited by high cost† and the high chemical reactivity of magnesium, which means that it is more susceptible to corrosion than aluminium. Neither of these considerations are important in one application, namely in the construction of spacecraft.

For parts of aircraft and spacecraft that get very hot, such as the skin of leading edges, titanium alloys have to be used. Titanium is a low-density refractory metal which has exceptional reactivity when molten. It is usually melted in vacuum by arc melting, using a water-cooled copper crucible so that the reactive molten titanium is separated from the copper by a layer of solid titanium.

9.4   Case 3. High-temperature materials

We include here a short section on high-temperature materials because, as a class of materials, they are rather different in fabrication methods than the more conventional alloys discussed so far.

Once we reach temperatures above a few hundred degrees Celsius, the microstructure of carbon steels will change, making their use impracticable. This can be overcome by using nickel–iron alloys, as nickel has a similar strength to iron (at thirty times the cost) but has an exceptional resistance to high temperatures and corrosion. Nickel alloys will do for parts that get to red heat, say inside a jet engine; however, at temperatures above this, the refractory (very high melting point) metals have to be used. These are metals like titanium, molybdenum, tantalum and tungsten. The latter two have melting points in the

---

†However, magnesium is abundant in seawater, so this may not apply in the future.

order of 3000 °C. These can be used for parts that reach white heat, like rocket nozzles or light-bulb filaments.

Although it is possible to fabricate them by vacuum arc melting, fabrication of parts with complex shapes from materials such as tantalum and tungsten is difficult. One very useful fabrication method uses the metal in the form that it retains after extraction from the ores: a fine powder. This method, which is given the name of powder metallurgy, is illustrated in Fig. 9.5.

The powder is mixed with a binder and packed into a flexible rubber mould. The mould is sealed with more rubber solution and placed in a holding cage. The whole lot is then placed in a hydrostatic press, where the oil pressure is pumped up to about 330 MN m^{-2}. This compacts the powder to such a high density that when it is removed from the press the rubber bag can be removed and the compact will have enough strength to remain intact. The compact is then placed in a furnace and heated at a high temperature (still well below the melting point)

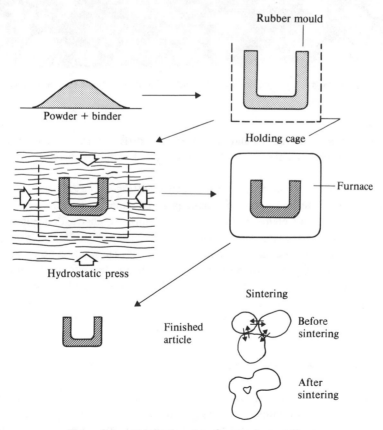

Figure 9.5   Fabrication steps for powder metallurgy.

Figure 9.6 Micrograph of the surface of a sintered alumina substrate (scanning electron microscope).

where the binder is driven off and there is enough diffusion to create an intimate bond between the particles where they touch one another. This latter process is called sintering.

To show that cost is only a prime consideration in cases where one has alternatives, this method is used to make large crucibles of iridium, one of the most expensive of all elements (several times more expensive than platinum). These are used for growing laser crystals at more than $2000\,^\circ$C (see Sec. 9.6). Only iridium has the necessary combination of very high melting point and low chemical reactivity (to prevent contamination of the molten laser material).

This method can equally easily be applied to ceramics, and this is done for producing shapes of magnesia (MgO) and alumina (Al_2O_3). The microstructure of the surface of sintered alumina produced as a substrate for electronic microcircuits is shown in Fig. 9.6.

9.5 Case 4. Electrical conductors and power transmission lines

The choice of metals to use as conductors of electrical current is extremely limited. One requires minimum electrical resistance so that energy is not wasted in generating heat. If one looks at the variation of electrical resistivity with composition for copper–nickel alloys (Fig. 9.7(*b*)), one can see that the addition of a second metal to create an alloy will greatly increase the electrical resistivity. The reason for this is that the distortion of the crystal lattice by the substitutional atoms increases the scattering of the electrons because more electrons will momentarily have energies in the band gap. The effects of the distortion in the lattice can also be seen in the variation in yield stress with composition (Fig. 9.7(*a*)). In fact, the solid solution Cu–Ni alloys which have high strength due to solid solution hardening are very useful for making resistors, furnace windings, and heater elements where electrical energy is deliberately used to generate heat. The extremely good resistance of these alloys to corrosion and high temperatures is a great asset for these applications, and also for other uses in high-temperature and/or corrosive environments such as for tubes in desalination plant and marine applications.

One is limited to only two metals for practical electrical conductors: pure copper and pure aluminium. The other good electrical conductors (silver, platinum, and gold) are far too expensive for large-scale applications.

The copper that is normally used, for power transmission lines, for example, is electrolytically refined 'tough pitch' copper. This contains 0.04 per cent oxygen as copper oxide and so has slightly higher resistivity than copper that is refined in oxygen-free conditions (oxygen-free high-conductivity copper). The latter is very expensive to produce and so is only used for special applications.

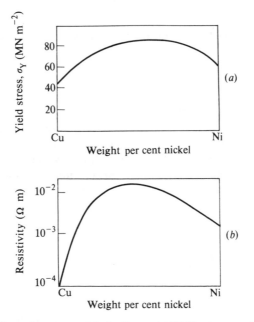

Figure 9.7 Properties of copper–nickel alloys. (*a*) Yield stress against composition; (*b*) resistivity against composition.

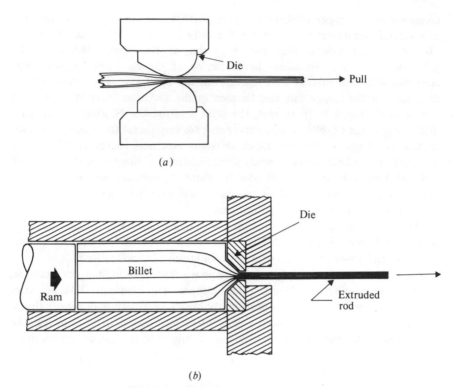

(a)

(b)

Figure 9.8   Two method of extrusion. (a) Wire drawing; (b) pressing of a billet (usually billet is pre-heated).

However, most of the oxygen can be removed by adding phosphorus to the melt (phosphorus-deoxidized copper). The copper is cast into bars and hot-rolled through shaped rollers to form rod which is then drawn through a series of dies (Fig. 9.8(a)) of successively smaller diameter to form wire.

A production method that is often cheaper than the traditional rolling and drawing route is extrusion of a billet. With this technique, a cast billet (often preheated) is placed in a press with a die at one end. A ram then compresses the billet, which is forced through the die. This achieves a tremendous reduction in area, equivalent to many rolling and drawing stages, in one blow. This is illustrated in Fig. 9.8(b). Seamless tube and cable covering can be made by extrusion. The latter method can be applied to either lead for corrosion protection (undersea cable) or plastic as an insulating cover. Extrusion can only be used for the ductile metals such as copper and aluminium, where it is a very useful technique for producing complicated sections such as those needed for window frames. Extrusion is also used for production of the solders (in wire or rod form) that are used for making low-resistivity joints in electrical circuits. The most common solder is soft solder, Pb–Sn, which is a eutectic system with a eutectic point at 72 per cent Sn and 183 °C. For higher-temperature applications, silver solders are often used based on the Ag–Cu eutectic, which has a eutectic point at 41 per cent Cu and 772 °C; this is a hard solder or brazing alloy. Solders have to alloy with the metals to be joined, so there is a large range of brazing alloys depending on the brazing temperature and the metals to be joined. A micrograph of a brazing alloy eutectic structure has already been shown in Fig. 7.12.

We have strayed rather far from the materials requirements for power transmission lines, but at least we have established that pure (or as pure as cost

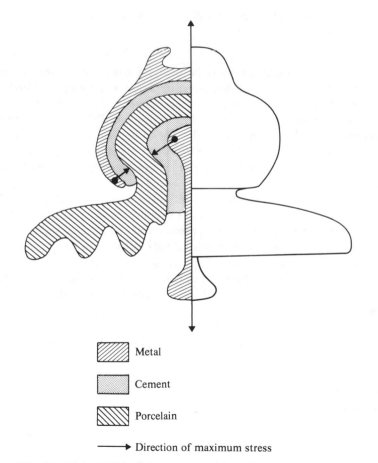

▨ Metal

▨ Cement

▨ Porcelain

⟶ Direction of maximum stress

Figure 9.9 Porcelain insulator for power transmission lines (left-hand side shows a cross-section).

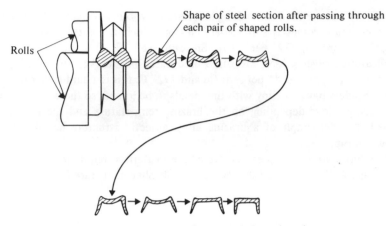

Figure 9.10   Stages in rolling a steel channel section.

will allow) copper or aluminium must be used for conductors. There are two other elements to power transmission lines: insulators and a support structure.

As the power loss is proportional to the square of the current times the resistance, the current must be kept as low as possible. This implies that very high voltages must be used, and so the line has to be well isolated from earth. Insulators have to be built capable not only of withstanding the electrical field without breakdown, but also of bearing the load of hundreds of yards of heavy copper wire.†

The necessary isolation is obtained by using a special porcelain shape to give a long surface path (Fig. 9.9) with a good glaze to eliminate any porocity that may trap water, as this would cause electrical breakdown. The glaze may also be very slightly conducting to even out the electrical field distribution along the surface in order to prevent charge building up, as this would lead to breakdown. Porcelain is weak in tension, so the insulator is cunningly designed so that, although the whole unit is in tension, the porcelain is only subject to compressive forces.

To support the lines, we need a tower of some sort and a steel fabrication is almost invariably used as it is strong, light, and can be made in prefabricated sections for easy erection.

Steel structures are usually fabricated from channel sections that are produced from rod by rolling in a succession of shaped rolls, as shown in Fig. 9.10, where production of a rolled steel joist (RSJ) is illustrated.

---

† Aluminium wire would not be much lighter as to have an equivalent resistance it would have to be of larger diameter than the copper wire.

9.6 Case 5. Pure single crystals

Many special ceramics are needed in single-crystal form. The prime example is silicon for semiconductor integrated circuits, where impurity levels of less than one part in 10^{13} are required. These semiconductor crystals are amongst the purest materials known to man.

Silicon of high purity cannot be grown by solidification in a crucible because a surface will be contaminated and strained by contact of the molten silicon with the crucible surface (for example, SiO_2 crucibles will contaminate the silicon with oxygen). Pure single crystals are grown by using a single crystal called a seed held in a mechanical grip. Crystal growing takes part in a sealed system with an inert atmosphere. The seed is lowered slowly to the surface of molten silicon in a large (inert) crucible (Fig. 9.11(a)). Silicon atoms start to attach themselves to the seed, which acts as a single nucleus, and so the seed starts to grow retaining the original crystal structure. The seed is slowly drawn out of the melt as the crystal grows and is also rotated to stir the melt and even out any temperature variations. Impurities from the crucible will be rejected from the growing solid (segregation).

Crystals grown by this method are very pure but need to be purified further before they can be used for manufacture of electronic circuits. The required purity is achieved by zone refining (Fig. 9.11(b)). This depends on the impurity lowering the liquidus and solidus in the way shown by the schematic fraction of

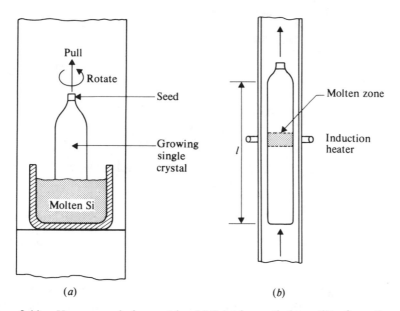

Figure 9.11   Very pure single crystals. (a) Crystal growth by pulling from the melt; (b) zone refining.

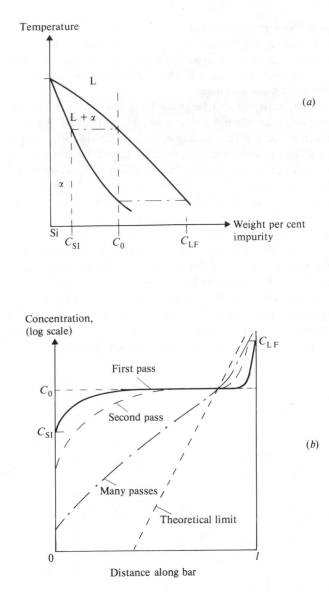

Figure 9.12 Zone refining. (*a*) Silicon–impurity phase diagram very close to pure silicon; (*b*) variation of impurity concentration with distance along the bar. C_0, starting concentration of impurity; C_{SI}, concentration for the first solid to form; C_{LF}, concentration for the last part to solidify.

the silicon–impurity phase diagram in Fig. 9.12(*a*). A molten zone is created by heating (by induced electrical currents or electron bombardment) at one end of the crystal whilst it is supported vertically in an inert atmosphere by means of grips at the ends. Surface tension holds the molten zone in place. The crystal is then slowly pulled upwards so that the molten zone travels along the crystal from one end to the other. If the initial concentration of impurity is C_0, then, as the zone moves and the bit that melted first starts to solidify, segregation will take place and the solid will have a lower concentration of impurity given by C_{SI}. Segregation will carry on as more silicon solidifies and the zone moves

Figure 9.13   Electron beam float zone refining of titanium.

down the bar; however, as more and more impurity is accumulated by the molten zone, the solid that forms will have more and more impurity incorporated.

A steady state is reached when the composition of the solid reaches C_0; at this point as much impurity is arriving in the liquid zone at one end as goes out at the other. There is no change from then on until the zone reaches the very end of the bar, where the heat is switched off and the extra load of impurity is left in the last bit to solidify, which will therefore have the composition C_{LF}. The distribution of impurity with distance along the bar is shown in Fig. 9.12(b). If this process is now repeated, segregation will once more operate, the initial part of the bar will get a little purer, the point at which a steady state is reached will get even further along the bar, and the last little bit will get even more impure. After many passes, one can reach very low impurity concentrations indeed at the 'good' end of the bar. The impure end is cut off and discarded. Another advantage of this technique is that any imperfections or strain in the crystal structure is removed by the passage of the zone; in fact, this method is often used for growth of single crystals from polycrystalline bars. The zone refining of a polycrystalline bar of titanium is illustrated in Fig. 9.13. This takes place in vacuum and heating is by electron bombardment from the circular filament.

Further reading

There are very many books that will take you further into the subjects covered here. Usually, there is an emphasis either towards the solid-state physics or materials science aspects of engineering solids. Four books that have been useful in the preparation of this book are

1. Hutchinson, T. S., and D. C. Baird, *The Physics of Engineering Solids*, 2nd ed., John Wiley, New York, 1968.

This book has a solid-state physics bias, but a fair amount of metallurgy is included. It is pitched at a level that is probably just right as a 'follow-on' from this book.

2. Flimn, R. A., and P. K. Trojan, *Engineering Materials and Their Applications*, Houghton Mifflin, Boston, 1975.

As the title implies, this has a very strong materials bias, with only a very cursory glance at physics. It is quite detailed on the steels and has many useful lists of properties, but in other areas is rather brief.

3. Kittel, C., *Introduction to Solid State Physics*, 4th ed., John Wiley, New York, 1971.

This is a book meant for physicists, and the non-physicist may find it hard going, but it is well written with good introductions to each section and is well illustrated.

4. Wert, C. A., and R. M. Thomson, *Physics of Solids*, McGraw-Hill, New York, 1964.

This covers most topics covered in this book in greater detail and so is a good 'follow-on'.

Index